DRONES
in Digital Agriculture

THE AUTHORS

Dr. K.S. Subramanian, Director of Research, Tamil Nadu Agricultural University, Coimbatore, India and awarded Ph.D. at the University of Ottawa, Canada, utilizing one of the most prestigious Canadian Commonwealth Scholarships (1993-1998). He has twenty two years of research and teaching experiences in TNAU and acquired expertise in the fields of nanotechnology in agriculture and soil biology. **Dr. K.S. Subramanian** is the Founder Head of the Department of Nano Science & Technology, developed research work for Agricultural Nanotechnology, Principal Investigator of several International and National Projects worth over **Rs. 30 Crores** (5 million USD) funded by Global Affairs Canada (GAC), International Development Research Center (IDRC), DST Nano Mission and ICAR.

Dr. S. Pazhanivelan, Professor and Head, Department of Remote Sensing and GIS, TNAU, Coimbatore, completed his Ph.D., in Agronomy during 1997. He has twenty two years of professional experience, specialization in Remote Sensing, Disaster monitoring, crop insurances and data science. He has completed 14 research projects including "Remote Sensing based Information for Crop Insurance in emerging Economies (RIICE) funded by Rice Consortium including *GIZ*. He has developed TNAU – RIICE technology of Remote sensing-based crop information for crop insurances, Smart sampling of Crop cutting Experiments using Remote sensing products and Mobile application for ground truth collection for crop monitoring and flood mapping.

Dr. R. Santhi, Professor (SS & AC) and Director, Directorate of Natural Resource Management, TNAU, Coimbatore has thirty three years of total experience in teaching, research, extension and administration. She has operated 24 funded projects and has expertise in Soil Test Crop Response – Integrated Plant Nutrition System (STCR-IPNS) Research for more than twenty five years and under her leadership, STCR – IPNS based fertilizer prescriptions for major Agricultural & Horticultural crops has been evolved and released as "TNAU Management Technology" and included in TNAU Crop Production Guide 2020.

Mr. G. Srinivasan, Ph.D. Scholar in Agronomy, is pursuing Ph.D. in the Department of Remote Sensing & GIS. He is undertaking research on area and yield estimations in cotton utilizing SAR data and DSSAT Model. His work is of great significance in the spread of cotton area in the State of Tamil Nadu as it is highly varying with the data presented by the crop cutting experiments of the Department of Statistics. Recently, he published a paper on "**Drone in Pest Management**" in Frontiers in Agronomy.

DRONES
in Digital Agriculture

— Authors —

Dr. K.S. Subramanian

Dr. S. Pazhanivelan

Dr. R. Santhi

Mr. G. Srinivasan

Tamil Nadu Agricultural University
Coimbatore – 641 003, Tamil Nadu

2022

Daya Publishing House®
A Division of

Astral International Pvt. Ltd.

New Delhi – 110 002

ISBN: 978-93-90371-67-9 (HB)

Published by : **Daya Publishing House®**
A Division of
Astral International Pvt. Ltd.
– ISO 9001:2015 Certified Company –
4736/23, Ansari Road, Darya Ganj
New Delhi-110 002
Ph. 011-43549197, 8130496929
E-mail: info@astralint.com
Website: www.astralint.com

Digitally Printed at : Replika Press Pvt. Ltd.

Dr. K. Alagusundaram, Ph.D.
Former Deputy Director General
(Agricultural Engineering)
Indian Council of Agricultural Research
New Delhi 110 012

भाकृअनुप

Preface

The Indian Agriculture has taken a paradigm shift from conventional to precision farming to tackle a wide array of field problems such as climate change, decline in organic matter content of soil, imbalanced crop nutrition, multi-nutrient deficiencies, frequent occurrence of invasive pests and diseases besides acute labour shortage.

In this context, **digital farming** becomes pertinent as it considers all complex issues in agriculture as a single technology package. Digital agriculture refers to the transformative farming practice wherein advanced tools and techniques such as geoinformatics, sensors, artificial intelligence and drones are exploited. These technologies are highly useful in predicting and quantifying exact quantity of inputs required for agriculture that eventually resulted in reduction in input use, improved productivity of crops while ensuring environmental safety.

Within the package of digital agriculture, drone is one of the emerging and timely technologies to reduce the drudgery of labour shortage and to overcome invasive pests like fall armyworm, spiraling whitefly in coconut, tea mosquito bugs in plantations and desert locusts in northern India etc. The new occurrence of devastating pests in the history of Indian Agriculture is least known and lot to happen in the years to come.

The Tamil Nadu Agricultural University, Coimbatore, is one of the best agricultural universities in the country and known for its innovation, begin to exploit drone technology to circumvent the labour shortage in the State of Tamil Nadu and timely detection of pests and diseases and smart delivery on inputs to promote productivity of crops. I take this opportunity to take congratulate Dr. K.S. Subramanian and his team in orchestrating a text book on "**Drones in Digital Agriculture**" which will be a guiding tool for students and scientists who have anchored research program in the niche area of drone technology. The book covers basics of drones, application of drones in seeding, fertilizer application, pesticides spraying and irrigation scheduling, plant health monitoring and risk assessment.

17/01/22

(K. Alagusundaram)

Contents

Abbreviations

ANN: Artificial Neural Network
AVHRR: Advanced Very High Resolution Radiometer
CCCI: Canopy Chlorophyll Content Index
°C: Celsius
CH: Canopy Height
CMI: Conventional Multispectral Indices
CNNs: Convolutional Neural Networks
CP: Crude Protein
CSMs: Crop Surface Models
CV: Coefficient of Variation
CWSI: Crop Water Stress Index
DAS: Days After Sowing
DBM: Dry Biomass
DEMs: Digital Elevation Models
DGCI: Dark Green Colour Index
DL: Deep Learning
DSM: Digital Surface Model
DTM: Digital Terrain Model
DVI: Difference Vegetation Index
ECa: Electricity Conductivity
ENDVI: Enhanced Normalized Difference Vegetation Index

EVI: Enhanced Vegetation Index
ExG: Excess Greenness Index
^{0}F: Fahrenheit
FBM: Fresh Biomass
FLAV: Flavonoid Index
GCPs: Ground Control Points
GCS: Ground Control Station
GDU: Cumulative Growing Degree Units
GIS: Geographical Information Systems
GNDVI: Green Normalized Difference Vegetation Index
GPS: Global Positioning System
gs -: Stomatal conductance
GSD: Ground Sample Distance
LAI: Leaf Area Index
LDM: Leaf Dry Matter
LiDAR: Light Detection and Ranging
LNA: Leaf Nitrogen Accumulation
LR: Simple Linear Regression
LST: Land Surface Temperature
Max: Maximum
Min: Minimum
MLR: Multiple Linear Regression
MODIS: Moderate Resolution Imaging Spectro-radiometer
MPa: Mega Pascal
MTVI: Modified Triangular Vegetation Index
NBI: Nitrogen Balance Index
NCC: Normalized Cross-Correlation
NDF: Neutral Detergent Fiber
NDI: Normalized Difference Index
NDRE: Normalized Difference Red Edge Index
NDVI: Normalized Difference Vegetation Index
NExG: Normalized Excess Green Index
NGRDI: Normalized Green Red Difference Index
NIR: Near Infrared
NNI: Crop Nitrogen Nutrition Index
OBIA: Object Based Image Analysis

OSAVI: Optimized Soil Adjusted Vegetation Index
PCA: Principal Component Analysis
PHT: Plant Height
PLS: Partial Least-Squares
PLSR: Partial Least Square Regression
PNA: Plant Nitrogen Accumulation
PRI: Photochemical Reflectance Index
RDI: Regulated deficit irrigation
RDVI: Renormalized Difference Vegetation Index
RE: Red Edge
RF: Random Forest
RGB: Red, Green, Blue
RMSE: Root Mean Square Error
RRMSE: Relative Root Mean Square Error
RS: Remote Sensing
RTK: Real Time Kinematic
RTNM: Real-Time Nitrogen Management
RVI: Ratio Vegetation Index
SAIL: Scattering of Arbitrarily Inclined Leaves
SCCCI: Simplified Canopy Chlorophyll Content Index
SD: Standard Deviation
SfM: Structure from Motion
SFR: Chlorophyll Fluorescence Index
SIFT: Scale Invariant Feature Transformation
SMA: Spectral Mixture Analysis
SMLR: Step-Wise Multiple Linear Regression
SONAR: Sound Navigation And Ranging
SPAD: N index values
SVIs: Spectral Vegetative Indices
SWP: Stem Water Potential
Ta: Air temperature
Tc: Canopy temperature
Tc: Crown temperature
2D: Two-Dimensional
3D: Three-Dimensional
TIR: Thermal Infrared

TM: Thematic Mapper imagery
TSS: Total Soluble Solids
TVDI: Temperature Vegetation Dryness Index
UAV-CS: UAV control system
UAVs: Unmanned Aerial Vehicles
VIs: Vegetation Indices
VRI: Variable Rate Irrigation
VRT: Variable-rate fertilization control technology
WDI: Water Deficit Index
WDRVI: Wide Dynamic Range Vegetation Index

Chapter 1

History of Drone

Over the years, agriculture has taken a paradigm shift from traditional farming to precision agriculture. Such transformation needs a series of changes in the production technologies from seed to seed till reaching the consumers. Tangible change in agricultural production practices is popularly referred as "**plow to plane**" that means agriculture being monitored remotely from the heaven. Despite the fact that the agriculture deals with cultivation of crops on the plow surface layer which is the skin of the earth, the activities and timely operations can be monitored through drones or satellite imageries. These technologies are widely used in the year 2020 in day today activities. Since there is acute labour shortage and the crops are being threatened by invasive pests and diseases, it is appropriate to introduce less labour intensive agriculture concomitantly blended with innovative technologies that can offer on-site solution and quick response time to undertake the corrective measures. In this context, drone technology is a powerful tool to accomplish the tasks set in precision agriculture. To get the comprehensive package about drone technology, this book chapter is devoted for exploring how exactly it evolved over the years in historic perspective.

The **drones** are unmanned miniaturized aircraft that can be remotely controlled or autonomously flown to undertake a specific task of monitoring the health status of crops or other purposes within a short span of time. Drones are being widely spoken these days in agriculture in lieu of labour shortage and as an effective monitoring tool for the police patrolling the people during the worldwide lockdown after the pandemic incidence of novel corona virus disease (COVID 19) during March – May 2020. In India, drones become popular during the corona curfew as the police and public have to maintain social distancing while ensuring safety of the people.

1.1. Genesis of Drone technology

If we look back in the history, drones are known to the world 150 years ago which truly unbelievable as it is frequently spoken in 2020 as a cutting edge technology. By definition **"drone"** is referred as an unmanned aerial vehicle (UAV). Chinese kites graced the skies to the first hot air balloon, unmanned flying craft came first before the risk of someone climbing on board occurred. One of the early users of unmanned aircraft was by the **Chinese General Zhuge Liang** (180–234 AD) who used paper balloons fitted with oil-burning lamps to heat the air; he then flew these over the enemy at night to make them think there was a divine force at work. Chinese are referred as pioneers in developing unmanned vehicles as part of military exercise.

One of the earliest UAV in the history of drones was seen in **1839,** when Austrian soldiers attacked the city of Venice with **unmanned balloons filled with explosives**. They used the balloons for attacking Venice on August 22, 1849. Some of these Austrian Balloons were successful, but many of them blew back and bombed the Austrians' own lines caused self destruction and thus they were discontinued. The invention of winged aircraft changed everything for manned and unmanned vehicles alike.

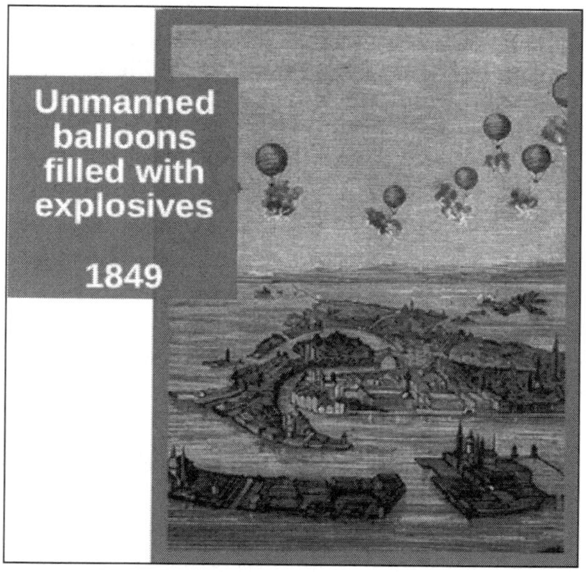

Fig. 1.1: (*Source*: https: //www.dronethusiast.com/history-of-drones/)

Few years later, in **1890,** the German aviation pioneer **Otto Lilienthal,** circa employed unmanned gliders as experimental test beds for main lifting wing designs and the development of lightweight aero structures. This caused several mishaps and advances were made in the designs to protect the pilot onboard. Although such approaches to remove people from the equation were used, the lack of a satisfactory method to affect control limited the use of these early unmanned aircraft. Early aviation developmental efforts quickly turned to the use of the first "test pilots" to fly these groundbreaking craft. Further advances beyond unmanned gliders proved painful as even pioneer **Lilienthal** was killed flying an experimental glider in 1896.

Historically, the unmanned aircraft often followed a consistent operational pattern, described today as the **three D's**, which stand for *dangerous*, *dirty*, and *dull* (Barnhart *et al.*, 2012). *Dangerous* being that someone is either trying to bring down the aircraft or where the life of the pilot may be at undue risk operationally. *Dirty* is where the environment may be contaminated by chemical, biological, or radiological hazards precluding human exposure. Finally, *dull* is where the task requires long hours in the air making manned flight fatiguing, stressful, and therefore not desirable.

1.2. Effective Control

The early stage of flights did not have effective control that lead to the mishaps. The Wright Brothers' success in flying the first airplane is more of a technical success story in solving the ability to control a piloted, heavier-than-air craft. The Wright Brothers are referred as the inventor of aircrafts. The first flight of the **Wright Brothers' famous Kitty Hawk flight** was flown in to the sky in the autumn of 1900. After 16 years later, the Great Britain developed the first pilotless winged aircraft: the **Ruston Proctor Aerial Target.** The Aerial Target was based on designs by Nikola Tesla and was controlled with radio control much like the drones of today but it was rudimentary. In **1915,** in the battle of Neuve Chapelle, the british used aerial imagery to capture more than 1500 sky view maps of the German trench fortification in the region.

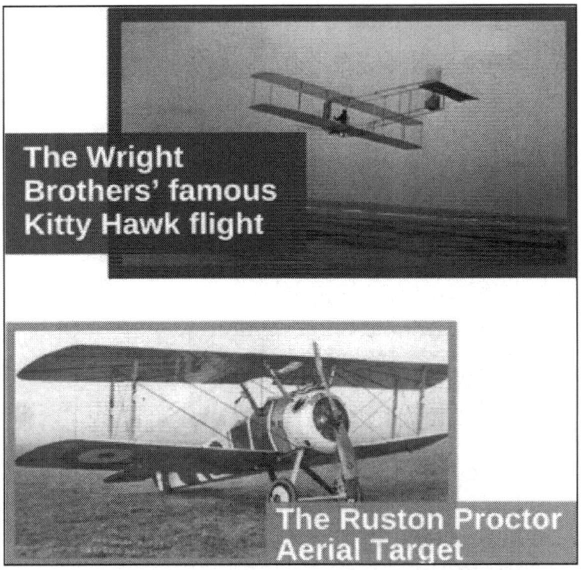

Fig. 1.2: (*Source*: https: //www.dronethusiast.com/history-of-drones/)

The British wanted to use the Aerial Target as a "**flying bomb**" which to counter the German Zeppelins besides using ground targets. After several failed prototype launches, the British military had chosen to scrap the project, believing that UAV had limited military potential. It was proved wrong. One year later in **1916,** Americans came up with an alternative as **Hewitt-Sperry Automatic Airplane.** It was tested and demonstrated successfully to US Army and such planes were mass produced and commissioned as "**Kettering Bug**". The Bug was a marvel of 1918 technology, but it was developed too late to be used in the war and was never deployed in combat.

Fig. 1.3: World War One: The Kettering Bug, World's First "Cruise Missile."

The UAV technology improved throughout World War II and into the Cold War as well. But they were largely seen as unreliable and expensive novelties – nowhere near as reliable as the exciting new innovations in manned aircraft like the Flying Fortress and the SR-71 Blackbird. While continue to develop UAV technology, USA navy has experimented a radio-controlled air craft resulting in the creation of the **Curtis N2C-2 drone** in **1937.** During WWII, Reginald Denny created the first remote controlled aircraft called the **Radioplane OQ-2.** This was mass multiplied and supplied for the military and it was considered as one of the landmark events in the history of development of drone technology.

The UAVs gained popularity and many countries begin to use this technology as a protective shield. Despite UAVs were considered as too expensive toy to use, the technological advancement has changed the attitude. What we do know is that modern drone warfare began in earnest in 1982, when Israel coordinated the use of battlefield UAVs alongside manned aircraft to wipe out the Syrian fleet with very minimal losses. The Israelis Air Force successfully combated Syrian Air Force in **1982** using **Battlefield UAVs** and manned aircraft to destroy Syrian aircrafts with minimal loss. The Israeli Air Force used military drones to recon the enemy's position, to jam communications, and to act as decoys that would prevent the loss of pilot life. The USA anchored Pioneer UAV program to fulfill the requirement of cheaper ones for fleet operations. In **1986,** a new drone was created RQ2 Pioneer in a joint project between US and Israel which largely used for reconnaissance survey.

In 1990, miniature and micro UAVs introduced and in **2000** USA deployed the predator done in Afghanistan in searching of Osma Bin Laden. Though the entire episode of events of development of drones is purely for military purposes, people begin use it for other purposes as it has access to the inaccessible areas and reaching the unreached in a short span of time. In **2014,** Amazon started using it for the shipment and delivery of customer goods and the realtors use it for shooting their properties for promotional activities.

The UAVs have been developed in a wide variety of shapes, sizes, configurations, and characteristics. The UAVs were used for different purposes and it began with military. The drones were used as an effective tool for surveillance during the military exercise. The use of UAVs has begun as early as 1900's, and the technology in its early stages was used by the military due in World War I and expanded by World War II (IJBE, 2013). Other than military applications, non-military applications of UAVs are relatively small but growing rapidly. The civil applications include firefighting, police patrolling, civil disturbances, crime detection and reconnaissance survey of natural calamities, border security forces, traffic surveillance and management, and scientific research in archaeological prospecting. In 1980's, the Government of India started using it for dusting pesticides that attracted queries from environmentalists on the safety of such practices to humans and animals welfare. India used drones for aerial seeding in forests as a part of reforestation programs and as a tool to access to reach the unreached destinations and hostile environmental conditions.

The use of UAVs has proliferated in other applications to relevant to agriculture such as aerial photography and imaging of crop fields to assist with crop health assessment and monitoring. Despite the fact that remote sensing data are collected using high-altitude flights from piloted aircrafts, UAVs were employed to fly at low altitudes to acquire the remotely sensed data. The UAVs are commonly exploited for the purpose of low altitude remote sensing (LARS). They can be either miniaturized fixed-wing airplanes or rotary-winged helicopters. The rotary wing helicopters are cheaper, slower speed, low ceiling altitude, light weight, low payload weight capability, and short endurance (Huang, 2010). In addition to these types, remote controlled (RC) kites, balloons, gliders, and motorized parafoils have been used for agricultural LARS imaging as well (Pudelko *et al.,* 2012). The UAVs for LARS are mostly fueled with gasoline or methanol and latest ones are being electric powered with rechargeable batteries or solar power. Since, chargeable batteries have limited power; they can be used only for short endurance. On the other hand, solar power can potentially provide longer endurance (Herwitz *et al.,* 2004). The UAVs used agricultural LARS is usually a short run and return back after capturing the images. In case of pesticide spraying UAVs, they require high payload weight requirement as it has to deliver the pesticides for a specific duration, quantity, time and height *etc.* This would require long duration flight endurance (Huang *et al.,* 2009).

Conventionally, the entire agricultural fields treated as uniform and the agronomic practices such as planting, fertilizer application, pesticide sprays, irrigation and harvesting is done same way without considering the variability within the field. Conversely, the precision agriculture segregates heterogeneous field into homogeneous domains and management of soil, nutrition, and crop stress considering local field needs are taken care accordingly. In the past ten years, advancement in electronics, Global Positioning Systems (GPS), Geographic Information Systems (GIS), and remote sensing have enabled remarkable changes in developing standard operational protocols for achieving precision agriculture. Remote sensing, especially from piloted airborne systems, has provided the basis of creating field prescriptions for variable crop seeding, fertilizer/pesticide/harvesting aid application, and irrigation. The agricultural uses of UAVs are more focused on low altitude operation over small fields for site-specific precision farming practice.

This is one of the key components of precision agriculture wherein inputs are applied at a right time, right place and right quantity utilizing the innovative approaches. In order to overcome the acute shortage of labours, drone technology is being exploited. The drones also frequently referred as unmanned aerial vehicles (UAVs) that developed and fine tuned to suit precision agriculture. Large scale spray of pesticides was taken up using piloted aircraft that carry a spraying to cover hundreds of hectares of crop fields in a short period. This caused lots of pollution related issues due to the fact that a specific pesticide is sprayed lavishly without discriminating crops or stages or threshold level of pests or diseases at the ground level. This necessitated going for UAVs that are miniaturized one and its operation can be controlled to suit small and marginal lands especially in India.

Recently, drones were employed to undertake spray in extensive maize grown areas in Tamil Nadu when the crop was threatened by an invasive insect pest **fall army worm** (*Spodeoptera furgiperda*). This pest causes extensive damage in a short span on time. Used. With maize cultivation getting heavily affected by invasive pest Fall Armyworm, the Tamil Nadu government had embarked on large-scale ground spraying of bio-pesticides/chemicals in the entire area of maize cultivation in the state. In October 2019, the state agriculture department has engaged the services of Madras Institute of Technology of Anna University to avail drones for aerial spraying of pesticides on a pilot basis in Perambalur district which is one of the hotspots for the invasive insect pest. Two sprays were done at an interval of 15-20 days and 40-45 days in 86,656 hectares at a cost of 5,500 per hectare. The state agriculture department on Wednesday unveiled the massive exercise at a cost of 47.66 crore. This was on the heels of announcing a compensatory package of 186.25 crore to the maize growers across the state. Polyphagous lepidopteron, an exotic pest which has a wide host range of 80 plant species, is causing major damage to the crop. Similar exercise was also performed in Tuticorin District where 1000 ha of maize fields were sprayed out of 15,000 ha in the districts. The loss of maize grains due to fall army worm was estimated as 3-4 lakh tones in 2019. Due to the close interventions of scientists of Tamil Nadu Agricultural University, Coimbatore, and the use of innovative technologies like drones have helped the State to protect the crop from Fall Army Worm during 2020.

On the other hand, use of drone technology for the large scale insecticide spray was objected in Telungana State in India. While extensive usage of drones to spray pesticide on agricultural fields is being reported from different parts of Telangana, the Ministry of Agriculture and Farmers' Welfare, in a reply to city-based activist Donthi Narasimha Reddy, said it was illegal as per the Insecticides Act of 1968. The Ministry also said that no permission or approval has been granted by the Central Insecticides Board (CIB) for spraying of pesticides using drones. This comes as a shock as a city-based drone startup has been spraying pesticides in Telangana as well as Andhra Pradesh and as per reports, it has sprayed pesticide using drones in over 5,000 acres in both the States. It may be mentioned here that the State government has also been planning to use drones for spraying of insecticides in the city for mosquito control.

Reddy, in his letter to the CIB, wrote, "I wish to bring to your notice the increased usage of drones for agro-chemical spraying, especially highly-hazardous pesticides We are also aware that farmers, and sprayers, ignore and/or cannot follow safety precautions under adverse weather and wind conditions." He further explained that the drift of spraying can take the fine hazardous chemical beyond the range of application. Reddy also pointed out, "You are aware that Insecticide Act does not allow aerial spraying, in general. It was done long before drones have been conceived as an idea and technology. Presently, we need to expand the specific clause that restricts aerial spraying to include drones and unmanned machines. Drones and unmanned machines can be hazardous tools for spraying hazardous chemicals, in many ways." Voicing his concern, Reddy also asked the CIB to issue an advisory to all the concerned regulatory bodies. "Drone manufacturers, suppliers and the entire network have to be cautioned against using drone for aerial spraying of agro-chemicals. Strict monitoring has to be enabled through State Agricultural Departments," the activist urged in his letter.

These pros and cons of drone technology tend indicate that the technology is powerful for rapid action and measures to manage the fast spreading invasive insect pests. This requires lots of basic information on the functioning of drones, optimization of various parameters for effective delivery of pesticides, selection of a right type of drones for agriculture besides economics of drones and regulatory guidelines. This book is an maiden attempt in India to the agricultural professionals to be aware of the technology and relate how best it can be employed to achieve digital farming.

Chapter 2

Why do we Need UAV in Indian Agriculture?

2.1. Genesis of Agriculture in India

India is one of the ancient civilizations where Indians have practiced agriculture even as early as Vedic period presumably started from c.1500 BC and ended in c.500 BC that corresponded with the last phase of the Chalcolithic period and Iron Age in India. The possible sites stretched from north-western parts of India to the entire alluvial of the river Ganges (Roy, 2009). Indians were responsible for the domestication of rice, millets, sugarcane and other crops. Though they practiced agriculture time immemorial, the yield levels of many of the crops were extremely low and were sufficient to feed one-third of the current population (1350 million in 2020) which was 439 million in 1961. The Indian agriculture growth has phenomenally increased ever since green revolution was in place in the country in 1960's. Despite the fact that the country was begging bowl waiting to get food stuffs from other countries like USA and Canada to nourish its population.

The country was facing two years of severe drought in 1965 and 1966 convinced that India should reform its agricultural policy and no longer rely on foreign aid and imports for food security. India adopted significant policy reforms focused on the goal of food grain self-sufficiency. The 'Green revolution" was anchored in the country by a renowned agricultural scientist **Prof. MS Swaminathan** and the then policy maker **Sri C. Subramanian** to adopt high yielding varieties of rice and wheat, disease resistant wheat varieties in combination with better farming knowledge to improve productivity. Consequently, the food grain production has increased more than quadrupled while the Indian population has tripled in the past sixty years ever since green revolution was introduced. The food grain production increased from **442 million tonnes** in 1960 to the all time high of **283 million tonnes** in 2020 and such a phenomenal grain growth is associated with the development of high yielding varieties, increased fertilizer consumption, irrigation and improved production technologies. The Indian buffer stock is currently 58.5 million tonnes in 2020 which is the highest indicating a strong food reserve in the country to face any challenges of extreme weather conditions.

Though we had achieved self-sufficiency in food grain production through green revolution, the associated ill-effects needs to be overcome through the evergreen revolution. The Father of the Green Revolution in India **Prof. M S Swaminathan** has introduced the word 'evergreen revolution' in 1993 which implies productivity improvement in perpetuity without ecological and social harm. The evergreen revolution involves the integration of ecological principles in technology development and dissemination. The major problems associated with the Green Revolution are related to environmental factors like depletion and pollution of groundwater, soil erosion and loss of biodiversity. It is these deficiencies that can be remedied through the 'evergreen revolution' pathway developed that was introduced in 1960's. Therefore 50 years later we should work for a transition from the green to an 'evergreen revolution', leading to an enhancement in productivity in perpetuity without ecological harm. He further stated that to achieve the goal of food security by ending hunger and to promote sustainable agriculture, it is important that in the field of social protection as well as the scientific measures needed for achieving food and nutrition security, we should move from the green to an 'evergreen revolution' approach.

The Prime Minister Narendra Modi had called for an "evergreen revolution" to enable the country to meet the challenge faced by the agriculture sector (The Hindu, May 19, 2017). He emphasized the need for moving from the concept of "food security" to "nutritional security" for which advanced scientific and technological interventions are required. We have to produce more food from less and less amount of land, labour and water. Thus there is scope of introducing nanotechnology for agriculture to meet the challenges ahead in the transformation of green revolution to evergreen revolution.

2.2. Need for Drones in Agriculture

Agricultural scientists are facing a wide spectrum of challenges in crop production system such as crop yield stagnation, declining organic matter, multi-nutrient deficiencies, climate change, shrinking arable land and water availability, resistance to GMOs and shortage of labour. Indian agriculture feels the pain of fatigue of green revolution. Despite the resounding success in grain growth, there are several constraints emerged in the past six decades since the inception of green revolution era.

- The imbalanced fertilization is primarily attributed to indiscriminate use of nitrogenous fertilizers and exclusion other macro- and micronutrients. The optimal NPK fertilizer ratio of 4: 2: 1 is ideal for crop productivity while the current ratio is being maintained at 10: 2.7: 1 in India. A nitrogenous fertilizer particularly urea is heavily subsidized by the government and thus its application is more obvious than other nutrients. Excessive N fertilizer use affects groundwater and also causes eutrophication in aquatic ecosystems. Drones may be useful to deliver customized fertilizers carrying macro and micronutrients to achieve balanced crop nutrition.
- Less fertilizer response ratio in the irrigated areas of the country, in other words, more amounts of fertilizers is required to produce the same quantity of grain output. For instance, 27 kg NPK ha^{-1} was required to produce one tonne of grain in 1970 while the same level of production can be achieved by 109 kg NPK ha^{-1} in 2008 (Biswas and Sharma 2008). The soil fertilizers have limitations in improving nutrient use efficiencies. Foliar feeding of fertilization using drones may be an alternative strategy.
- In order to achieve a target of 300 million tones of food grains and to feed the burgeoning population of 1.4 billion in the year 2025, the country will require 45 million tonnes of nutrients as against a current consumption level of 27 million tonnes. The extent of multi-nutrient deficiencies are alarmingly increasing year by year which is closely associated with a crop loss of nearly 25- 30 per cent.
- The extent of nutrient deficiencies in the country is in the order of 89, 80, 50, 41, 49 and 33 per cent for N, P, K, S, Zn and B, respectively. The data from long-term fertilizer experiments unequivocally demonstrated that conjunctive use of organic and inorganic is essential to sustain soil health. But the availability of organic manures is becoming scarce as a result of urbanization and reduction in animal wealth.
- Frequent occurrence of devastating invasive pests like fall army worm in maize, rugose spiraling white fly in coconut and diseases like blast in rice, red rot in sugarcane and viral diseases in horticultural crops that require quick delivery of pesticides to manage the crop
- Climate change is yet another serious concern. Eratic rainfall, frequent occurrence of drought, melting polar ice cap, temperature rise (0.4 to 4.0°C in the past century that coincides with increase in CO_2 concentration from 280 ppm in 1900 to 430 ppm in 2020) and declining biodiversity.
- There is an alarming problem of acute labour shortage in the country causing havoc in agricultural growth as many farmers quitting farming. In order to overcome scientists are working on complete mechanization for major field crops.
- This lead to evolution of the concept of smart farming wherein advanced tools and techniques such as sensors in agriculture, artificial intelligence

in crop health monitoring, drones for detection of pests and diseases, risk assessment and timely delivery of agricultural inputs and robotics in agriculture

Considering the current scenario in Indian agriculture, there is an urgent need to know the use of drone technology in the entire value chain in agriculture. Despite it is fairly recent in India, other countries have used drone technology for spraying agricultural inputs for various crops and agro-ecological regions. Aerial application, usually called aerial crop spraying, involves spraying crops with fertilizers, pesticides, fungicides, and other crop protection materials using agricultural aircraft (Lan *et al.,* 2010). The countries that possess developed agricultural aviation mainly include the United States, Russia, Australia, Canada, Brazil, Japan, and Korea. In thinly populated areas of the United States, manned fixed-wing aircraft is the most popular form of agricultural aviation (Xue and Lan, 2013). The developmental mode of agricultural aviation in Russia, Australia, Canada, and Brazil shares the same characteristics as the United State, the main types of which are manned fixed wing aircraft and rotor helicopters. In contrast, there are many mountains and very little arable land in Japan and Korea, which are unsuitable for the use of manned fixed wing aircraft. Agricultural aviation in these areas are thus dominated by small unmanned aerial vehicles (UAV). Compared to conventional agricultural aircraft, UAVs do not require a special airport and have advantages, such as good mobility (Chen and Lu, 2012; Bae and Koo, 2013). UAVs are also more adaptable for spraying at low altitudes due to geographical restrictions (Zhang *et al.,* 2014; Lan *et al.,* 2008).

Chapter 3

Components and Types of Drones

3.1. Components of UAV Systems

There are four basic components in drones encompassing UAVs, Ground Control Station (GCS), UAV control system (UAV-CS) and Sensors for data acquisition. Among the four, UAV is the heart of the drone system being operated automatically or remotely.

3.1.1. UAVs

The unmanned aerial vehicles (UAVs) are one of the major components in UAV system that have no operator on their spindle but operated either autonomously or remotely. The type of UAV to be exploited is depending on nature of operations.

3.1.2. Ground Control Station (GCS)

The ground control station is purely a computer controlling and managing the movement of UAVs. It is a computer that either communicates with the UAV Control System or controls and monitors the UAV directly. The GCS monitors information related to the flight of the UAV. The user has the ability to receive data relevant to the flight of the aircraft, but also data recorded by the sensors that support the flight (*i.e.*, ground-based sensors or sensors embedded in the aircraft). In addition, the GCS contains the software required for the processing of data acquired by the UAV and the extraction of the information needed by the system operator for the crop monitoring.

3.1.3. UAV Control System (UAV CS)

The UAV Control System is used to control the UAV. This is similar to air control system in airports. It can be either a two-way data link, such as a remote control, or a built-in computer (usually with a built in GPS). The UAV CS includes the flight control system and/or the autopilot system, which controls the operation of the UAV. This system receives and processes data from the autopilot or flight control system for the proper operation of the UAV. It usually contains sensors to monitor the flight properties, such as sensors for measuring distance from ground, air force, *etc.* The control system has the ability to process information from sensors to correct any problems that may arise, and to communicate with the GCC wireless and in real time by sending and receiving the necessary information.

3.1.4. Sensors for Data Acquisition

In order to take imageries or data of the point of interest, cameras or sensors are attached. Depending on the type or accuracy of the data, the pixels differ. They are cameras intended to collect the information needed. The next section provides a detailed presentation of possible ways of collecting information and the technologies exploited. In the case that UAVs are not intended to collect information but are used for another purpose, such as spraying, the sensors are replaced with the necessary components. The UAVs are multi-functional vehicles and the components like sensors or cameras or payload of spray formulations can be fixed as per the requirement.

3.2. Types of Drones

The drones are classified based on types (fixed-wing, multirotor, *etc.*), the degree of autonomy, the size and weight, and the power source. This categorization is very important to decide what type of drones can be employed for different purposes within agriculture. In agriculture, we have different types of crops (field crops, fruits, vegetables, tree crops *etc.*), applications (sowing, aerial seeding, planting, spraying *etc.*) and other purposes (health monitoring, risk assessment). The flight height, speed, duration, payloads and sensors will vary depending on the purpose or conditions. The types of drones will be decided depending on the siuations in which they are going to be employed.

Classification of Drones

The majority of existing drones can be classified into two types namely fixed-wing systems and multirotor systems. Others are hybrid systems and ornithopters.

Fixed-wing is defined as aircraft that use fixed, static wings in combination with forward airspeed to generate lift. Examples for fixed wing include traditional airplanes, kites that are attached to the surface and hang gliders or paragliders. Raven is a good example for fixed wing type.

Multirotor Systems are a subset of rotorcraft. Rotorcraft is defined as an aircraft that use rotary wings to generate lift. A popular example of a rotorcraft is the traditional helicopter. Multirotor drone is the widely used Phantom drone made by the Chinese company DJI. Multirotor drones do not need a landing strip, make less noise than their fixed wing counterparts and can hover in the air. Fixed-wing drones can fly faster and are more suitable for long distances than their multirotor counterparts. These characteristics determine which of these drone types to use for a specific application.

Hybrid systems have characteristics of both multirotor and fixed-wing systems. Hybrid quadcopter is an example for hybrid system. **Ornithopter** type is neither fixed wing nor multi-coloured and is far less frequent. These drones fly by mimicking wing motions of insects or birds. Most of these ornithopters are scaled to the birds and are not widely used in practice. Examples of ornithopters include the Delfly explorer that mimics a dragonfly.

In agriculture, multi-rotor and ornithopter are widely used for assessing the health status, time of harvest and yield of the crops. In addition, the same set of UAVs can be used for spraying of pesticides, nutrients and other inputs.

3.3. Design Characteristics of Drones

The UAVs are widely used in precision agriculture and the types of UAVs can be further divided into five basic categories, based on their design characteristics. As per the designs, UAVs can be categorized into 5 types namely (1) fixed wing, (2) rotary wing, (3) blimps, (4) flapping wing and (5) parafoil wing.

3.3.1. Fixed-wing

The fixed wing UAVs require a runway to take off from the ground or a catapult. They have high endurance as well as the ability to fly at high speeds and capable of covering larger areas on each flight and can carry more payload. However, they are more expensive than the other types. Solar-powered UAVs offer significantly increased flight times because they exploit and store the sun's energy during the day. This is the reason that they are preferred for long-endurance operations. The fixed wing type will be useful for aerial seeding, forest fire containment and rescue operations where the flying time is long and high endurance is required.

3.3.2. Rotary-wing

The rotary-wing UAVs, also called rotorcrafts or Vertical Take-Off and Landing (VTOL), offer the advantages of steady flying at one place while keeping the maneuverability attribute. These features are useful for many different types of missions. However, they fly at low speed and short endurance. They are generally the most widely used UAVs in all kinds of applications, but especially in Precision Agriculture. One reason for this is the fact that they present lower cost compared to the other types of UAVs. In addition, this type of UAVs is suitable when the monitored crops are not very large, which is usually the case.

Rotary wing type is sub categorized into (a) unmanned helicopter and (b) Multi-color. The unmanned helicopter includes main and tail rotors such as conventional helicopters. Only 4 per cent is used under this category. Multi-rotor includes rotary-wing UAVs with four or more rotors (quadcopter, hexacopter, octocopter, *etc.*). These aircraft are generally more stable in flight than unmanned

(a)

(b) Cai *et al.*, 2014

(c) Mozgeris *et al.*, 2018

(d) Palossi *et al.*, 2017

(e) Cai *et al.*, 2014

(f) Hunt *et al.*, 2014

Figure 3.1: (a) Fixed Wing (Department of Remote Sensing and GIS, TNAU, Coimbatore); (b) Helicopter (Hornet Maxi); (c) Quad Copter (Department of Remote Sensing and GIS, TNAU, Coimbatore); (d) Blimp; (e) Flapping-wing (SmartBird); and (f) Parafoil-wing (Tetracam).

helicopters. Overall, 72 per cent of the works used this type. The rotary wing type is used in pesticide sprays and plant health monitoring besides risk assessment. Rotary wing type is widely used in agriculture for a wide range of applications.

3.3.3. Blimps

This type of UAV is lighter than air, has high endurance, flies at low speeds and is generally larger in size compared to the other types. Their manufacturing characteristics will allow them to remain in the air even in the event of a total loss of power, while being considered relatively safe in the event of a collision. Usually, they are not used in Precision Agriculture applications.

3.3.4. Flapping Wing

These UAVs are very small and they have flexible, shaped little wings inspired by the way birds and insects fly. They are **not often used** in Precision Agriculture as they require high energy consumption due to their size.

3.3.5. Parafoil-wing

Usually aircrafts of this type have one or more propellers at the back in order to control the course of their flight, but at the same time for harnessing the power of the air to fly without consuming much energy. They are also capable of carrying a larger payload. They are not usually exploited for precision agriculture applications. Only 2 per cent of the works analyzed use this type.

Among the various categories of drones, rotary-wing aircraft are preferred as they are easy to operate, slower speeds, ability to maneuver and relatively low cost. These advantages provide greater opportunities for collecting information from crops through imaging, which is the main use of UAVs in vegetation monitoring. In cases where the monitoring area is relatively large, fixed-wing aircraft are preferred, which enable the monitoring of the entire area in a short time. Among the 5 types of drone, rotary wing type is widely used in precision agriculture as they are cheap and versatile applications.

3.4. Level of Autonomy

Drones are unmanned vehicles and in the absence of a pilot, drones always have a certain level of autonomy. An important distinction within the concept of autonomy is the difference between **automatic** and **autonomous** systems. An automatic system is a fully preprogrammed system that can perform a preprogrammed assignment on its own. Automation also includes aspects like automatic flight stabilization. Autonomous systems, on the other hand, can deal with unexpected situations by using a preprogrammed rule set to help them make choices. Automatic systems cannot exercise this 'freedom of choice.

3.5. Size and Weight

Other important characteristics of a drone are its size and weight. Clarke (2014) distinguishes large drones and small drones, but divides the small drones in multiple subcategories. Clarke suggested that the lower weight limit of large drones is **150 kg for fixed-wing drones** and **100 kg for multirotor drones**. Many countries distinguish large and small (or light and heavy) drones. For instance, the

Dutch Human Environment and Transport Inspectorate (ILT) make a distinction between light drones (< 150 kgs) and heavy drones (> 150 kgs).

The development of drones is currently focused on making smaller and lighter drones for the general public. Large drones are mainly used for military purposes. Therefore, a shift can be observed from large drones to smaller drones. This calls for changing the reference categories and the category parameters. Therefore, they suggest using the term **large drones** for fixed-wing drones between 20 and 150 kg and multirotor drones between 25 and 100 kg. **Small drones** are fixed-wing drones up to 20 kg and multirotor drones up to 25 kg. Within the category of small drones, they suggest to use a subcategory of mini drones. Mini drones can vary in weight from several grams up to several kilograms. These mini drones are mainly suitable for indoor applications and recreational applications.

3.6. Power Source

There are four main energy sources namely traditional airplane fuel, battery cells, fuel cells, and solar cells.

Airplane fuel (kerosene) is mainly used in large fixed-wing drones such as military Predator drone. This drone is used a lot by the US army and can be equipped with a number of different sensors, but also with rockets and other types of ammunition.

Battery cells are mainly used in smaller multirotor drones. These drones are short range and require less operating time than drones using kerosene. These drones are often for recreational use, making it more practical for the drone to run on a rechargeable battery cell. An example of such a drone is the above-mentioned Phantom drone.

A **fuel cell** is an electrochemical device that converts chemical energy from fuel directly into electrical energy. Because of the lack of conversions in thermic and mechanical energy, this conversion is efficient and environment friendly. For example, the Stalker drone which uses a fuel cell has a flight time of 8 h instead of 2 h.

Drones using **solar cells** are rare in the current drone industry. Drones using solar cells are mainly fixed-wing drones. Solar cells are suitable for small ornithopters. Among various energy sources, battery operated UAVs are commonly used in precision agriculture. Chargeable batteries are also available in the market to operate UAVs. Solar powered UAVs are yet to gain momentum in the market due to practical difficulties such as sunshine hours, rusting of solar panels and technical snags.

3.7. Drone Models

a. Delft Explorer

It is an ornithopter drone that flies like a dragonfly and is being developed by Delft University of Technology in the Netherlands. The drone can take-off and fly fully autonomous within a closed environment. It can avoid obstacles by using two cameras. The drone has a weight of 20 g and can currently operate for only nine minutes because of the size and weight constraints of the battery.

b. Hubsan x4 Drone

The Hubsan x4 is a small multirotor drone developed by the Chinese company Hubsan. This mini drone is fairly simple in design and operation. It has four rotors and can be operated with a controller. Some models of the x4 drone come with a built-in camera for making pictures and recording video. The drone has a weight of 30 g, a radius of around 100 m and can operate for 7 min with a fully charged battery.

c. Parrot AR Drone

The Parrot is a drone mainly built for recreational purposes. It has a multirotor system that can be controlled by a smartphone or tablet. The drone can operate for 12–18 min and weighs about 400 g. Its speed is about 18 km/h and it has a range of about 50 m. The drone has two cameras, Bluetooth and WiFi technology and uses GPS-waypoints to fly a preprogrammed route. The Parrot is similar to the below-mentioned Phantom, both in applications and functions. Besides film and photography software, the drone is also equipped with gaming software, making its emphasis on recreation more clear. The gaming aspect includes a racing game and augmented reality driven shooter games in which a real-world environment is augmented by computer-generated graphics and/or sound.

d. DJI Phantom

The Phantom drone is a multirotor drone with four rotors and is mainly built for recreational purposes. The drone comes with a camera and can be controlled using a smartphone or a WiFi controller. The smartphone can also control the camera to move and make pictures or record video. The Phantom can fly at around 54 km/h and it can operate for about 25 min. Just by programming the flight altitude and certain waypoints the drone can take-off, land, make recordings, and return automatically.

e. Raven

The Raven is a fixed-wing drone developed in 2002. The drone was originally developed for the US Army but is frequently used by many other countries as well, making it one of the most widely used drones in the world at this moment. The main purpose of the Raven is surveillance and it can be controlled remotely or preprogrammed for autonomous operation. The Raven has a width of 1.4 m, weighs about 2 kg, and can stay operational for 60–90 min within a range of 10 km. It is equipped with an optic and an infrared camera. Like regular model airplanes, the Raven can be launched by throwing it in the air. It lands by gliding toward a preprogrammed landing site and can compensate for the impact when hitting the ground by falling apart.

f. ScanEagle

The ScanEagle is a fixed-wing drone dating from 2004 and is mainly used as a surveillance tool. It is equipped with an optic and/or infrared camera and can operate for over 20 h. It is 3.1 m in width, 1.2 m in length, weighs 18 kg and has a cruising velocity of 89 km/h. The drone can be launched by pneumatic pressure and it can land with a skyhook system, plucking it out of the air. Therefore, a landing strip is not necessary. Contrary to most fixed-wing drones, the ScanEagle needs little space to take-off or land. In this section, a number of core characteristics of drones were described to determine the main differences between drones and their technical properties.

3.8. Types of Payloads and their Applications

All kinds of payloads can be attached to drones, the only restrictions are usually the weight and size of payloads. Most drones are equipped with cameras by its manufacturer. Other payloads can be ordered at drone manufacturers, but drone users also can attach payloads to their drone themselves. The payloads include sensors and other types of payloads.

Sensors

The weight, model, and energy source of a drone are major factors influencing its maximum altitude, flight duration, flight range, and maximum payload. An important category of payloads are sensors. Cameras and microphones are the major payloads for drones and often come standard when buying a drone. Cameras can be regular or infrared. Such cameras may enable night vision and heat sensing. Other sensors include biological sensors that can trace microorganisms, chemical sensors ('sniffers') that can measure chemical compositions and traces of particular chemical substances including radioactive particles and meteorological sensors that can measure wind, temperature, humidity, *etc.*

Drones are used for a wide array of applications in **defense**, police, traffic regulation, tracking criminals cheaper than aerial photography from manned aircraft and also cheaper than satellite imaging. Drones attached with particle sensors can be used for detecting the **emission** of particulates. Concentrations and emission rates of sulfur oxides, nitrogen oxides, and ammonia can be measured. Other sensors can measure light, sound, and radiation. These applications of drones may contribute to the **environment** and are also less polluting than manned aircraft. Drones may also help monitoring illegal waste dump and transport of toxic waste. When particular animals are provided with RFID tags, drones can track migrations, biodiversity, poaching, and habitats. Images created by drones may also be useful for estimating animal populations and tracking their behavior.

Drones with sensors are currently used in **agriculture**, for instance, for monitoring crop growth, estimating biomass, checking for weeds and plant diseases, and evaluating the quality and level of water. The use of drones in border surveillance is particularly useful in vast areas and areas that are difficult to reach or access.

3.9. European System of Standardization

Drones require radio systems to allow communication between the drone and the pilot. European Aviation Safety Agency (EASA) directive 216 is only applicable for drones with a weight above 150 kg and only for control and non-payload communications. In the European Economic Area the radio equipment on board drones up to 150 kg therefore need to comply with the essential requirements of the R and TTE69 and EMC70 directives for command and control communications. All pure payload communications in drones need to comply with these directives. The R and TTE directive will be replaced by the RED71 directive by June 2016. Manufacturers and importers have the responsibility for compliance of their drones before placing them on the market. If the drone complies with the essential requirements a CE marking has to be affixed to the drone or eventually to the packaging or the accompanying documents. Furthermore, a declaration of conformity has to be published.

3.10. UAV Data Acquisition

The UAVs attached with sensors are becoming powerful sensing systems that complement or supplement to the IoT-based techniques. Sensors capture high-spatial and temporal resolution images that can assist in monitoring many different characteristics of the vegetation. A variety of different types of sensors can be used in an agricultural UAV depending on the different crop parameters that should be monitored. The details are given in tables indicating how they are different under various situations such as sowing, herbicide application, irrigation, crop health monitoring *etc.* Selection criteria for sensors includes low weight, the low energy consumption and the small size that eventually resulted in capturing of high resolution images.

Modern commercial on-board sensors complying with the above restrictions that are used for precision agriculture, mainly belong to the following four types namely (1) Visible light sensors (RGB), (2) Multispectral sensors, (3) Hyperspectral sensors and (4) Thermal sensors

1. Visible light sensors (RGB)

The RGB (red, Green, Blue) sensors are the most frequently used sensors by UAV systems for Precision Agriculture applications. They are relatively low cost compared to the other types and can acquire high resolution images. In addition, they are easy to use and operate and they are lightweight. In addition, the information acquired requires simple processing. The images can be acquired in different conditions, on both sunny and cloudy days, but a specific time frame is required based on weather conditions to avoid inadequate or excessive exposure of the image. Considering the drawbacks of these sensors, the main disadvantage is the fact that they are inadequate for analyzing a lot of vegetation parameters that require spectral information in the non-visible spectrum. They are commonly used in tandem with the other types of sensors.

2 and 3. Multispectral or Hyperspectral Imaging Sensors

The UAVs can acquire information about the vegetation's spectral absorption and reflection on several bands. Spectral information can be significantly helpful in assessing a lot of biological and physical characteristics of the crops. For example, unhealthy parts of the crops can be discriminated in an image, as visible radiation in the red channel is absorbed by chlorophyll; while near infrared (NIR) radiation is strongly reflected. Thus, even if it is not yet visible in the red channel, it can be identified by the information in the NIR channel. Spectral information can be used to calculate several vegetation indices and monitor several crop characteristics based on them. Multispectral and hyperspectral sensors are frequently used, despite their higher costs. However, a drawback of these sensors arises from the fact that it is required to apply more complex pre-processing methods in order to extract useful information from the captured images. The pre-processing procedure of spectral images often contains the radiometric calibration, geometric correction, image fusion and image enhancement. The main difference between multispectral and hyperspectral sensors is the number of bands (or channels) that each sensor can capture and the width of the bands. Multispectral sensors capture 5–12 channels while hyperspectral images can usually capture hundreds or thousands bands, but in a narrower bandwidth. Although in the recent works studied multispectral sensors are used a lot more frequently than hyperspectral because of their lower cost, hyperspectral technology seems to have a lot of potential and is considered as the future trend for crop phenotyping research.

4. Thermal Infrared Sensors

These sensors capture information about the temperature of the objects and generate images displaying them based on this information and not their visible properties. Thermal cameras use infrared sensors and an optical lens to receive infrared energy. All objects warmer than absolute zero (- 273°C/- 459°F) emit infrared radiation at specific wavelengths (LWIR and MWIR bands) in an amount proportional to their temperature. Hence, thermal cameras focus and detect the radiation in these wavelengths and usually translate it into a grayscale image for the heat representation. Many thermal imaging sensors can also generate colored images. These images often show warmer objects as yellow and cooler objects as blue. This type of sensors is used for very specific applications (*e.g.*, irrigation management). As a result, they are not frequently used in precision agriculture applications of UAV systems that focus on monitoring other characteristics of the crops. In the image acquisition by UAVs, it is typical to acquire several overlapping images of the crops. In the most cases they capture both front and side overlapping images. This is desired as the overlapping images can be used for the construction

Figure 3.2: Examples of Sensors Used by UAVs for PA: (a) Thermal sensor (Maimaitijiang *et al.*, 2017); (b) RGB sensor (Hassanein *et al.*, 2018); (c) Multi-spectral sensor (Department of Remote Sensing and GIS, TNAU, Coimbatore); and (d) Hyperspectral sensor (Bhandari *et al.*, 2018).

of 3D models and/or orthophotos of the crops, as the next section discusses. The rate of the overlap depends on the type of the application. The front overlap usually ranges 60–95 per cent while the side overlap ranges 40–95 per cent to generate three-dimensional models and 25–40 per cent for other uses.

In addition, the altitude of the UAV flights varies according to the application and spatial accuracy of the information we want to collect. The distance between the target (crop) to be visualized and the UAV plays an important role in determining the detail of the information acquired. This is something that depends on both the sensors and the resolution they offer. In the majority of cases, depending on the purpose of the application, the spatial resolution of the photographs is between

0.5 cm/pixel and 10 cm/pixel. Quite often, RGB sensors are modified to acquire information about the radiation in other bands too, usually the Near Infrared (NIR) or the Red Edge (RE) band. This approach is observed when the stakeholders want to avoid the higher costs of buying multispectral cameras. This is achieved by replacing one of the original optical filters with one that enables the perception of near-infrared channel, resulting often in a hybrid (*e.g.*, NIR-RGB) sensor. The visible channel that is no longer captured by the modified RGB sensor is often captured by using another embedded RGB sensor.

In addition to the above types of sensors, Laser scanners are a well-established technology used extensively for environmental sciences, however they are mostly used for terrestrial scanning. Each sensor type can monitor different characteristics of the vegetation, such as the color and texture of vegetation or the geometric outline of the crops. In addition, some sensors can measure the radiation in certain wavelengths. The data acquired by these sensors can be further processed to monitor plant biomass, vegetation health, soil moisture and other important crop characteristics at the different growth stages.

3.11. UAS Data Processing

The most common features that can be monitored with UAV-based remote sensing for Precision Agriculture.

Crop Features

The remote sensing is widely used in precision agriculture. The plant parameters used include Vegetation biomass (Nasi *et al.,* 2018; Maimaitijiang *et al.,* 2017), Nitrogen status (Ballester *et al.,* 2017). Moisture content (Bhandari *et al.,* 2018), Vegetation color (Wan *et al.,* 2018; Varela *et al.,* 2018), Spectral behavior of chlorophyll (Quebrajo *et al.,* 2018) Temperature (Pascuzzi *et al.,* 2018), Spatial position of an object (De Castro *et al.,* 2018; Gao *et al.,* 2018), Size and shape of different elements and plants Vegetation indices (Marino *et al.,* 2018; Marcial-Pablo *et al.,* 2019).

Soil

In precision agriculture, soil parameters such as Moisture content(Jorge *et al.,* 2019; Sobayo *et al.,* 2018), Temperature and Electrical conductivity (Krizova *et al.,* 2018) are used for the determination of health status.

With the use of specialized sensors, UAVs can acquire information for various features of the cultivated field. However, as mentioned above, there is still no standardized workflow or well established techniques to follow for analyzing and visualizing the information acquired. The most commonly used image processing methods y to analyze UAV imagery for Precision Agriculture

Photogrammetry Techniques

Photogrammetry assists in accurate reconstruction of a scene or an object from several overlapping pictures. This technique can process the 2D data and establish the geometric relationships between the different images and the object(s), obtaining 3D models. To construct the 3D models, photogrammetry requires at least two overlapping images of the same scene and/or object(s), captured from different points of view. These kind of techniques can be used for extracting three-dimensional digital surface or terrain models (Ziliani *et al.,* 2018; Torres-Sánchez *et al.,* 2018; De Castro *et al.,* 2018) and/or orthophotos (Sa *et al.,* 2018).

UAV low-altitude data acquisition enables the construction of 3D models with a much higher spatial resolution compared to other remote sensing technologies (such as satellites). However, the collection of many images is required to have information for the entire field under study. Thus, in most cases, it is necessary to collect many overlapping images to construct Digital Elevation Models (DEMs) of the crops and/or create orthophotos (also referred to as orthomosaics). The 3D models and the orthophotos include information about the 3D characteristics of the crops based on the structure of the vegetation (*e.g.,* the vegetation height, the canopy, the density, *etc.*) and can be very useful for applications that can exploit only RGB imagery. The works reviewed showed that photogrammetric techniques are very commonly used in all types of applications as they are also required to create vegetation indices maps. In addition, the 3D information they include is very important and is often used in tandem with other techniques.

Machine Learning Methods

Machine Learning (ML) has been used to process the data acquired, for prediction and/or identification purposes, with great results in many domains, such as medical systems (Tsouros *et al.,* 2017; Bonotis *et al.,* 2019), marketing (Cui *et al.,* 2005), biology (Tarca *et al.,* 2007), *etc.* Machine learning techniques are often been applied in Precision Agriculture to exploit the information from the large amount of data acquired by the UAVs. ML is able to estimate some parameters regarding the crop growth rate, detect diseases or even to identify/discriminate objects in the images. Machine learning usage has increased a lot recently due to the fast advancements taking place especially in the deep learning field.

Vegetation Indices Calculation

Vegetation Indices (VIs) are one of the most popular products of remote sensing applications for Precision Agriculture. They use different mathematical combinations/transformations of at least two spectral bands of the electromagnetic spectrum, designed to maximize the contribution of the vegetation characteristics while minimizing the external confounding factors. They can deliver reliable spatial and temporal information about the agricultural crops monitored. In most cases, many VIs are extracted and used to draw conclusions. They can be calculated based on information of either each photograph individually or after the production of orthophotos depicting the whole crop. Calculating vegetation indices may serve in the identification of useful crop characteristics, such as biological and physical parameters of the vegetation. Since the processing of data may be time consuming, several software tools and techniques have been developed to enable faster data processing. The most commonly adopted software solutions in the works reviewed to support and accelerate the data analysis procedure are summarized in Table 2. In addition to the software tools referenced in the table, there are several other promising software tools that can assist in the data analysis process, such as Erdas Imagine (Leica Geosystems, 2004), eCognition (Baatz *et al.*, 2004), and PixelWrench 2 (Tetracam, 2011).

Most common software tools used in the literature for image processing.

Software Tool	**Description**
Adobe Photoshop (Wang *et al.*, 2018)	Applied to correct distortion/use of other image processing methods
Agisoft Photoscan (Fan *et al.*, 2018; Ziliani *et al.*, 2018)	Exploited for the construction of 3D models and orthomosaics. It also allows the calculation of vegetation indices
QGIS (Yonah *et al.*, 2018; Marino *et al.*, 2018)	Usually exploited for the calculation of the vegetation indices from multispectral data
MATLAB (Gnadinger *et al.*, 2017)	Applied mainly for the calculation of vegetation indices. It can also be exploited for other image processing methods
Pix4D (Raeva *et al.*, 2019; Khan *et al.*, 2018)	The most commonly used tool. It can be used for calculating VIs and/or constructing of 3D models and orthomosaics

3.12. Data Processing Techniques Used Agriculture

a. Photogrammetric Techniques

Photogrammetric techniques are mainly be applied for the construction of orthomosaics and/or Digital Elevation Models (DEMs) in order to exploit the 3D information regarding the vegetation. As mentioned above, this can be done by acquiring many overlapping images of the agricultural crops monitored. The advancements in computer vision and photogrammetric techniques have lead to many different techniques and algorithms that are able to match large numbers of overlapping images and detect common objects and scenes in them. The images are processed by applying aerial triangulation and adjusting camera orientation. Computer vision methods are used for matching the overlapping images and the common characteristics. To achieve object tracking as well as to identify the scale and the orientation of a particular image; in some cases, Ground Control Points (GCPs) are being used. GCPs distributed in the cultivated field can be identified within the overlapping images: (a) to link the images; and (b) to identify the coordinates of each image and its slope. However, recent advancements have made this redundant, as many techniques can be used without GCPs, with similar precision. The construction of the 3D Digital Elevation Models can provide to the producer information about the altitude of the earth surface, the natural and artificial objects/structures on the surface, the density of the vegetation, and their growth, among others.

There are two types of DEMs namely **Digital Terrain Model (DTM)** and Digital Surface Model (DSM). **DTM** represents the altitude of the surface of the Earth, *i.e.*, of the terrain. These models do not take into account either artificial or natural (*e.g.*, trees, vegetation, buildings) objects that exist in the field. DTMs just present the elevation of the bare Earth. Figure 3.3 shows a Digital Terrain Model from Ronchetti *et al.*, 2018.**DSM** represents the altitude of the surface that is first encountered by the remote sensing system (*i.e.*, when the aerial image captures the top of a building, tree, the vegetation *etc.*). Hence, the elevation model generated includes the elevation of the bare Earth along with artificial and natural objects that may exist in the field. The **DEMs** constructed can be exploited either for the extraction of 3D information directly or to construct orthomosaics of the crops. An orthoimage, orthomosaic or orthophoto is an aerial photograph that is orthorectified (*i.e.*, geometrically corrected). Thus, the scale of the constructed image is uniform. As a result, the final orthophoto has the same lack of distortion as a map. In contrast with a simple aerial image of a field, an orthophoto can be used to measure true distances as it contains the 3D characteristics of the crops. The use of photogrammetry techniques have the following procedure: create point cloud representations of the 3D surface and either combine all objects into a single Digital Elevation Model or use the DEMs to generate an orthophoto. The most commonly used set of algorithms for this purpose is Structure from Motion (SfM) (Westoby *et al.*, 2012).

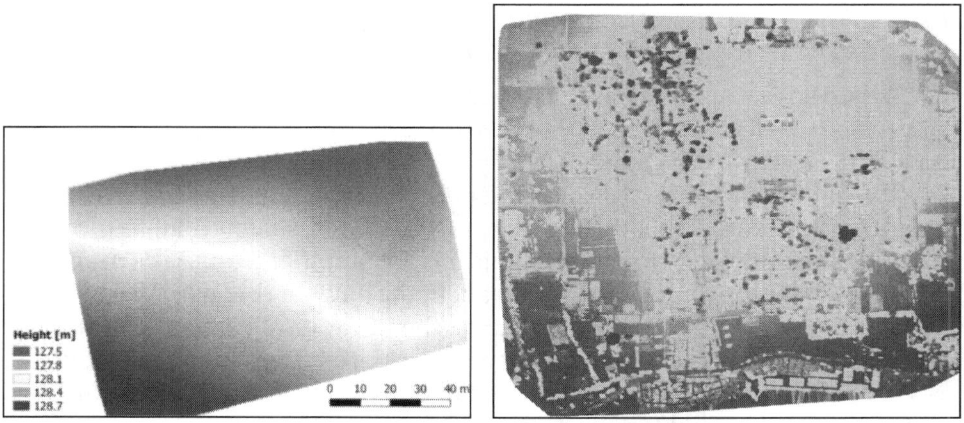

Figure 3.3: (a) Digital Terrain Model (b) Digital Surface Model of TNAU, Coimbatore

Photogrammetric techniques are commonly used in all types of applications as they are also required for constructing the vegetation indices maps. Photogrammetric techniques are used in the majority of recent works (93 per cent) to exploit and extract information regarding the 3D characteristics of the crops.

However, photogrammetric techniques are in most cases used to compliment other types of data processing methods.

b. Using Machine Learning

Machine Learning (ML) and Data Mining methods are widely used in precision agriculture to exploit the information acquired by the UAVs. Taking into account the large amounts of data collected from agricultural fields, machine learning can be applied to enhance the performance of UAV-based systems for precision agriculture, by extracting knowledge for several parameters of the vegetation. ML is used in many cases and for different purposes. Both unsupervised and supervised learning techniques are being exploited, via clustering, classification and regression methods. Regression methods are widely used in UAV applications for PA for a variety of purposes. Regression has been used to estimate spectral vegetation indices by analyzing data acquired from RGB images (Khan *et al.,* 2018) presenting generally good results. Additionally, regression has been used to examine the correlation of some vegetation indices with vegetation features such as nitrogen (Nasi *et al.,* 2018; Zheng *et al.,* 2018**;** Ballester *et al.,* 2017), leaf area index (Fan *et al.,* 2018; Maimaitijiang *et al.,* 2017), and biomass (Nasi *et al.,* 2018; Fan *et al.,* 2018). For this purpose, both linear (simple and multiple) regression and nonlinear regression methods have been used. In a comparison of different regression algorithms, for estimating leaf nitrogen content, Random Forests presents the best results among 14 algorithms, with the coefficient of determination (R^2) being up to 0.79. Regression methods are also used to predict crop water status, by using information derived from RGB, multispectral and thermal sensors (Romero *et al.,* 2018). In this context, Artificial Neural Networks present generally good results (R2 up to 0.87 by using and information from some spectral bands).

Classification methods are also very commonly used for weed mapping (Bah *et al.,* 2018; Franco *et al.,* 2018) and disease detection (Kerkech *et al.,* 2018; Montero *et al.,* 2018). The most popular and precise classification techniques are the Artificial Neural Networks (ANNs) family (Bah *et al.,* 2018) and the Random Forest algorithm (Wan *et al.,* 2018; Zheng *et al.,* 2018). These algorithms directly use the RGB colors, the intensity, spectral information or other features derived from the image acquired. In some cases, data about the neighborhood of each pixel are also considered. Apart from the above data, classification algorithms can also use vegetation indices as features in the model to achieve higher accuracy. In general, ANNs present higher accuracy, compared to other classification algorithms, that reaches up to 99 per cent for weed mapping in some cases (Dos Santos Ferreira *et al.,* 2017). The accuracy of the method depends on the type of the crops monitored (Bah *et al.,* 2018) as expected. Convolutional Neural Networks (CNNs), is among the most used family of algorithms. CNNs belong to deep learning algorithms that have been proved to be very effective in object detection in large datasets.

The use of **Deep Learning (DL)** in Precision Agriculture applications is a recent, modern and promising technique, having increasing popularity. Deep learning techniques extend typical ML by adding more complexity into the derived models. DL techniques transform the data using various functions that allow data representation in a hierarchical way, through several levels of abstraction. Advancements and applications of Deep Learning into other domains indicate its large potential. As indicated in (Liakos *et al.,* 2018; Kamilaris *et al.,* 2018), the use of Machine Learning and more particular Deep Learning will be even more widespread in the next years.

A very common application of Machine Learning methods in PA is Object Based Image Analysis (OBIA). The purpose of OBIA is to discriminate objects within agricultural images obtained from UAVs (De Castro *et al.,* 2018). In contrast to traditional pixel-based image classification, which classifies each pixel, OBIA groups small pixels together into vector objects. With the higher spatial resolution of UAV imagery, pixel-based classifications have become much less effective. That is due to the fact that the relationship between the pixel size and the object size has changed significantly. Therefore, object-oriented classification methods are increasing in popularity. They use segmentation methods to divide the image pixels into homogeneous segments/groups. Then, these segments/objects are arranged into classes based on their geometric, spectral, textural and other characteristics. OBIA is usually composed of two main steps:

1. Image segmentation
2. Feature extraction and classification

These methods are exploited to recognize and detect weeds or discriminate different species in the field. A detailed review of algorithms and challenges for OBIA, from a remote sensing perspective, was reported by Hossain and Chen, 2019.

c. Vegetation Indices

Vegetation Indices (VIs) have been widely used in remote sensing applications for Precision Agriculture. They are considered to be very effective for monitoring the growth and health of crops in qualitative and quantitative vegetation analysis (Mozgeris *et al.,* 2018; Huang *et al.,* 2018). Vegetation Indices are based on the absorption of electromagnetic radiation by the vegetation. They are mathematical transformations of the absorption and scattering in different bands of the electromagnetic spectrum. The reflectance in several bands is affected by parameters such as vegetation biochemical and physical properties, environmental effects, soil background properties, moisture content, *etc.* The understanding of the spectral behavior of the vegetation is fundamental to remote sensing applications to monitor various vegetation features (*e.g.,* biomass, nitrogen status (Nasi *et al.,* 2018), vegetation health (Kerkech *et al.,* 2018), *etc.*) It has been shown that certain VIs are significantly related with different parameters of the vegetation. Simple Vegetation Indices that can combine RGB information and some spectral bands such as NIR and RE have significantly improved the ability to detect green and healthy vegetation.

Several VIs have been developed, as different environments have their own complex characteristics, which needs to be taken into account when using a Vegetation Index. Thus, each VI has its own specific combination of the reflectance in different bands, in order to detect vegetation. Hence, its VI is suitable for specific uses. The main concept is to combine the reflections of different bands to decrease the "noise" from external factors (*e.g.*, sensors calibration, lighting, atmosphere, soil properties, *etc.*). For example, as mentioned in the previous section, visible radiation to the red is absorbed by the chlorophyll while the radiation in the NIR band is strongly reflected. In this way, vegetation can be discriminated by the soil in an image. In way, unhealthy vegetation can also be detected.

Vegetation Indices that are based on the radiation in the Red and NIR channels, such as the RVI or the NDVI index, are designed to increase the contrast between the vegetation and the soil. The relationship between the reflections of the two zones allows the elimination of disturbances by factors that affect the radiation of each zone in the same way. The effort to model the biophysical parameters of vegetation has led to the creation of several different vegetation indices (Wiegand *et al.,* 1991; Tanriverdi *et al.,* 2006; Bannari *et al.,* 1995; Xue *et al.,* 2017). The vegetation indices can be divided into two main categories:

1. Vegetation Indices based on multispectral or hyperspectral data. Most of the developed Vegetation Indices use multispectral and/or hyperspectral information that can combine several bands.
2. Vegetation Indices based on information from the visible spectrum. Several VIs in the visible spectrum have been developed and are widely used due to the high cost of multispectral and hyperspectral sensors.

A list of the most used vegetation indices is presented in Table 3. For the interpretation of the following formulas, the following abbreviations represent the reflection in the respective color or spectrum:

R: Red (620–670 nm)

G: Green (500–560 nm)

B: Blue (430–500 nm)

NIR: Near Infrared (720–1500 nm)

RE: Red Edge (670–720 nm)

Concerning the multispectral vegetation indices, one of the first well-known indices was **Ratio Vegetation Index** (RVI). This index enhances the contrast among vegetation and soil. However, it is sensitive to the optical properties of ground. The best known and most widely used vegetation index is the **Normalized Difference Vegetation Index** (NDVI), which is the evolution of RVI and is calculated by the visible and near infrared light reflected from the vegetation. Unhealthy or sparse vegetation reflects more visible light and less near infrared light, making it easy to monitor the growth and health of many agricultural crops. It is based on absorption in Red due to chlorophyll and reflectance in NIR. RVI and NDVI are calculated as shown in Table 3.

Vegetation Indices Derived from Multispectral Information

Ratio Vegetation Index (RVI)	NIR/R
Normalized Difference Vegetation Index (NDVI)	NIR – R/NIR + R
Normalized Difference Red Edge Index (NDRE)	NIR – RE/NIR + RE
Green NDVI (GNDVI)	NIR – G/NIR + G

RGB-based Vegetation Indices

Excess Greenness Index (E x G)	2 x G – R - B
Normalized Difference Index NDI	G - R/G + R

Several other VIs have been developed based on NDVI. NDRE uses the method of NDVI to normalize the ratio of NIR radiation with Red Edge (RE) radiation. The same applies for GNDVI with NIR and Green (G) bands.

Figure 3.4 shows examples of crop maps constructed from information of spectral VIs (NDVI and NDRE) in different growth stages. We can see that the difference in the maps while the vegetation grows is quite clear.

Focusing on the VIs extracted from RGB images, Excess Greenness Index (ExG) and Normalized Difference Index (NDI) are the most used indices. ExG is based on the assumption that plants display a clear high degree of greenness, and soil is the only background element. Thus, it is calculated by doubling the radiation in the Green channel minus the radiation in Red and Blue channels. NDI was proposed to separate plants from soil and residue background images, using only green and red channels. Although the VIs that use information in the visible light can be useful for crop monitoring, they cannot provide information for several parameters of the vegetation and also they are sensitive to working environment properties such as the atmosphere, lighting, *etc.* Hyperspectral remote sensing is expected to be the future trend in crop monitoring and this is mainly because it allows the development of new bands combination of vegetation indices. In many cases, it has been proved that hyperspectral vegetation indices are less sensitive to saturation, change in viewing/lighting geometry, and atmospheric contamination. The combination of new bands can eliminate noise from the working environment and in the same time exploit the information of certain bands and extract information for more biophysical features of the vegetation.

Despite the fact that UAVs can be applied in the entire episode of precision agriculture, major part focuses on sensors for monitoring the growth of crops and the processing methods exploiting UAV imagery.

- ☆ **Photogrammetry techniques** are used to construct Orthophotos or DEMs from the overlapping images acquired. They are used in most of the applications to create vegetation maps considering several

(a) NDVI (Normalized Difference Vegetation Index) of the cotton field (TNAU, Coimbatore).

(b) NDRE (Normalized Difference Red-edge index) of the cotton field (TNAU, Coimbatore).

Figure 3.4: Vegetation Indices Maps of Crops in different Growth Stages (Raeva *et al.,* 2019).

characteristics; however, they are also used standalone mainly when only RGB sensors are available, to exploit the 3D characteristics of the vegetation.

- ☆ **Machine Learning** methods can be used to monitor several different characteristics of the crops. They can exploit RGB and/or multispectral/ hyperspectral images. A very frequent technique exploited in the literature is the Object Based Image Analysis.
- ☆ **Vegetation Indices** use combinations of the reflection of several bands obtained by RGB or spectral sensors. They are proved to be most effective when multispectral or hyperspectral information is used. They have been used in many recent studies. They have been proved to be very effective in monitoring various parameters of the crops, by using different combinations of spectral bands. used to exploit UAV-acquired information for the same type of application.
- ☆ **Monitoring the health of the crops**: Disease detection is mainly applied through temporal analysis of the vegetation's growth. However, multispectral and hyperspectral sensors seem to have great potential in this context. An approach using multispectral sensing technologies could reliably deliver Vegetation Indices, discriminating healthy and unhealthy portions of the vegetation. In such an approach, the farmer could locate the "weak" corps field areas and take timely decisions to save the parts of the crop production that seem unhealthy.
- ☆ **Weed mapping**, we can see that some studies use machine learning techniques to detect weeds, mainly OBIA, while others use the 3D characteristics of the crops though DEMs, VIs or other processing methods. Weed detection with UAVs based on object-based image analysis seems to be at an advanced stage and can be used for site-specific weed management. We can see that machine learning techniques are exploited in 62.50 per cent of the applications that used RGB sensors and in all the applications with multispectral sensors.
- ☆ **Growth monitoring**, many methods are really promising but it seems that their integration in a standardized workflow can improve its applicability and efficiency. Growth monitoring and yield prediction is most commonly performed with the use of RGB (30 per cent) and multispectral (59 per cent) sensors. The information acquired is exploited for estimating the density of production and the biomass. The findings of this review appoint that combining information coming from Vegetation Indices with the 3D characteristics of the crops coming from RGB images can improve the accuracy of growth and yield estimation methods.
- ☆ **Irrigation management** is closer to a standard workflow, which involves the use of thermal and/or multispectral sensors to monitor the needs of water of different parts of the crops. Equipped with thermal sensors, the

Unmanned Aerial Vehicles are able to detect possible deficiencies in the irrigation of different sites. This information can be processed through photogrammetry techniques into a single high-resolution vegetation map, highlighting the stressed areas. A map in this context can be constructed though the use of some VIs when multispectral imagery is used. Thus, a map such as this can support the producers to apply Variable Rate Irrigation (VRI) applications

This chapter provided basic information on components of drones, types of drones, energy source of drones and payloads of drones. The payloads include sensors covering four types (visible, multi-spectral, hyper spectral and thermal infrared) and their applications in precision agriculture. The data processing for plant and soil parameters are also explained. In precision agriculture, how the data processing done using software tools and techniques is also elucidated. The techniques encompass photogrametry, machine learning, deep machine learning and sensors (multi-spectral and hyper spectral). Various indices such VI, NDVI, OBIA, DEMs and others are elaborated in order to provide the readers basic information on the drone technology that is applied in various spheres of agricultural sciences.

Chapter 4

Drones for Sowing

Seeds are prime input in agriculture deciding the productivity to the tune of 30-40 per cent. Improved crop varieties with good quality seeds are utmost important to sustain farm productivity. In India, more than 70 per cent of people directly or indirectly in agriculture which accounts for about 800 million people. In order to overcome the acute labour shortage across the country, agricultural scientists in the country are involved in design and fabricate machines to suit Indian conditions and to achieve complete mechanization in the country. Adding further complexity, in Indian farming activities closely coincide with monsoonic rains and the labour shortage gets aggravated during two peak cropping seasons namely *kharif* (June – July) and *rabi* (September – October). To win over the constraints, drones have been attempted for sowing and planting. Usage of drones is likely to reduce the labour requirement and enable timely sowing with the available moisture in the soil which very critical in rainfed farming. Since more than 60 per cent is rainfed farming, agricultural sustainability of country squarely depends on rainfed farming.

Diwate *et al.* (2018) have designed a drone for seed sowing and hypothetically explained that there is a possibility of reducing the labour requirement using drone-enabled seed sowing. The efficiency of sowing, germination, growth, yield and the economics of drone sowing is not presented. Very scantly information is available for droned aided sowing. However, extensive studies have been done in aerial seeding in forests particularly in Canada and Australia which serve as the base to build on to develop drones for sowing in Indian agriculture.

Seed germination and establishment of seedlings in forestry or natural ecosystems are equally challenging, as the mortality at the early stages often exceeding 95 per cent, are considered the most significant bottleneck in ecological restoration (James *et al.,* 2011; Hallett *et al.,* 2014). The seed-based ecological restoration projects has increased dramatically in recent decades, and it is now frequently undertaken in areas where traditional on-foot surveys of seedling emergence and early plant establishment can be challenging (*e.g.*, large post-mining landforms such as waste rock dumps and tailings storage facilities). Additionally, counting and classifying seedlings to species level can be time-consuming and extremely difficult in diverse seedling communities Stucky *et al.,* 1984; McDonald *et al.,* 1996; Hardwick *et al.,* 2004). As such, there is an increasingly urgent need for technological approaches offering restoration practitioners with fine-resolution, rapid and scalable plant-based monitoring solutions.

Among the suite of new technologies offering advances in ecological monitoring, unmanned aerial vehicles (UAVs) have significant potential for improving plant-based outcomes. The monitoring of vegetation growth and performance greatly benefits from UAV-based remote sensing (Duzgun *et al.,* 2011) mainly due to the scales at which monitoring must be undertaken. UAVs are considered the best option for remote sensing at scales up to 250 ha (Felderhof *et al.,* 2012), as they offer greater spatial resolution compared with satellite imagery(Berni *et al.,* 2009), significantly lower operational costs compared with manned aircraft (Ogden, 2013), and can be operated in weather conditions that would prevent both satellites and manned aircraft from gathering meaningful data (Hunt *et al.,* 2010; Koh *et al.,* 2012).Furthermore, remote sensing using UAVs offer greater repeatability and lower turnaround times than both satellites and manned aircraft (Nishar *et al.,* 2016). While alternative sensing platforms such as pole photogrammetry may offer utility in the monitoring of fixed and small-scale areas of interest (*e.g.*, vegetation monitoring quadrats), UAV-based remote sensing techniques have been successfully applied to the classification of adult plants (species of invasive trees and shrubs) in bogs and disturbed forest environments (Knoth *et al.,* 2013; Baena *et al.,* 2017; Lehmann *et al.,* 2017), as well as to monitor plant health in a variety of agricultural species (Berni *et al.,* 2009; Yue *et al.,* 2012; Calderón *et al.,* 2013; Candiago *et al.,* 2015). However, the utility of UAVs has been demonstrated almost exclusively in agricultural settings, and little translational research has been undertaken to apply this technology to plant-based monitoring in ecological restoration (Buters *et al.,* 2019). Buters *et al.* (2019) tested the utility of a small, commercially-available UAV to monitor seed germination and seedling establishment in restoration.

The monitoring of seeds and seedlings, which represent extremely vulnerable stages in the plant life cycle, is particularly challenging due to their diminutive size and lack of distinctive morphological characteristics. Counting and classifying seedlings to species level can be time-consuming and extremely difficult, and there is a need for technological approaches offering restoration practitioners with fine-resolution, rapid and scalable plant-based monitoring solutions. Buters *et al.* (2019) have exploited UAVs for seed and seedling monitoring by utilizing the combination of high-resolution sensors and low flight altitudes. This allows for the detection and monitoring of small objects, even in challenging terrain and in remote areas. They utilized the low-altitude UAV imageries and an automated object-based image analysis (OBIA) software to detect and count target seeds and seedlings from a matrix of non-target grasses across a variety of substrates reflective of local restoration substrates. Buters *et al.* (2019) have shown that the automated classification of target seeds and target seedlings were achieved with accuracies of exceeding 90 per cent and 80 per cent, respectively. The classification accuracy decreased with increasing flight altitude (*i.e.*, decreasing image resolution) and increasing background surface complexity (increasing percentage cover of non-target grasses and substrate surface texture). Results represent the first empirical evidence that small objects such as seeds and seedlings can be classified from complex ecological backgrounds using automated processes from UAV-imagery with high levels of accuracy. Buters *et al.* (2019) suggested that the novel application of UAV use in ecological monitoring offers restoration practitioners an excellent tool for rapid, reliable and non-destructive early restoration trajectory assessment.

The BioCarbon Engineering, a UK-based company backed by drone manufacturer Parrot, has been using the flying robots to plant trees on a massive scale. It has come up with a method of planting trees quickly and cheaply. Not only that, but those areas which are unviable or not fertile can also be converted into growing land. The drones, which were developed by an ex-NASA engineer, are designed to fire off pre-germinated seed pods into the ground. This machine has been made plant much faster than manual planting methods. The BioCarbon is closely working with non-profit organization Worldwide International Foundation intended to replant mangrove seedlings in Myanmar. The evidence of restoration was quite obvious by checking the grown up seedlings that have reached 20".

Fig. 4.1: (*Source*: https: //krishijagran.com/farm-mechanization/ drone-that-can-plant-10-000-trees-a-day/)

The drones can be employed for a series of operations leading to successful restoration of forest or natural vegetations. Firstly, drones can be used to scan the topography to create a 3D map. Then, the most efficient planting pattern for that area is calculated using algorithms. The seeds are fired at the rate of one per second into the ground by drones loaded with seeds. Sixty drone teams can plant 1 billion trees a year. The BioCarbon also tested this technology in reseeding mining sites in Dungog, Australia. The method is 10 times faster than manual seeding with a cost reduction by 80 per cent. There are additional advantages that include seeding in terrains, inaccessible areas.

The **Government of Norway** has set up a new fund in 2017 to kick-start investments in deforestation-free agriculture in countries that are working to reduce their forest and peat degradation. It will work in partnership with the Global Environment Facility, UN Environment Programme, the Sustainable Trade Initiative (IDH) and major food companies and environmental NGOs, and aims to protect over 5 million hectares of forests and peatlands by 2020, equivalent to the size of Costa Rica. It has been estimated that 74,000 to 94,000 KM^2 forest areas is lost every which is equivalent to 48,000 football fields lost every minute. If two workers are trained to operate fleet of 10 drones then planting of 400,000 trees per day can be accomplished. It has assumed that if this proves successful the drones could soon help to resuscitate degraded landscape around the world.

A **Canadian Tech Company** is aiming to plant 1 billion trees by 2028–and they plan on achieving their goal with drones. Toronto-based startup Flash Forest is using a fleet of aerial drones which has been designed to plant trees ten times faster than human planters. Since they began testing their prototype drones earlier this year, they have already managed to plant several thousand trees across Ontario. Using pre-germinated seed pods, the drones are capable of planting trees at just 50 cents per sapling–which is just one-fourth the cost of typical planting methods.

Thus, the **success stories** in Norway, Myanmar and Canada clearly demonstrated that drone technology is very powerful in terms of area coverage, less labour and rapid restoration of forests and natural vegetation. But, seeding in field crops is least known and one of the potential areas to be exploited in the near future. In field crops, there is a possibility that small sized seeds can be pelleted as a spherical to enable sowing of rounded seeds to get sown. There is a vast scope for agricultural scientists or seed technologists to work together with aeronautic engineers to optimize the parameters for effective seeding in a short span of time. In India, sowing is taken up spontaneously immediately after monsoon rains and the shortage of labour will be acute and UAVs could be a potential solution.

Chapter 5

Drones for Nutrient Management

Fertilizers play a pivotal role in sustaining farm productivity. After the green revolution era started in 1960's our food grain production has increased from 50 million tonnes to 283 million tonnes (2019) in India that closely synchronized with the use of fertilizers from 0.5 million tonnes to 27 million tonnes for the same period. Despite use of fertilizers has increased exponentially, there is a bundle of constraints faced by Indian farming includes imbalanced fertilization, decline in organic matter in soils, occurrence of multi-micronutrient deficiencies besides less fertilizer response ration. This necessitates rationalization and scientific and site specific nutrient application to ensure augmented productivity while safeguarding environmental safety. It is, therefore, inevitably in demand for realizing the sustainable development of modern agriculture (Liu *et al.,* 2015; Zhenhua *et al.,* 2011).

5.1. Variable-rate Fertilization

The variable rate fertilization is the foundation of precision agriculture that intends to meet the 'green planting' demands of modern agriculture with demand driven input and balanced fertilization (Yuan *et al.,* 2010). The variable-rate technology encopasses **"3S"** technology (*i.e.,* remote sensing [RS], Geographical Information Systems [GIS], and the Global Positioning System [GPS]) to acquire real-time crop growth and nutrient gain–loss information as the basis for on-demand and variable-rate fertilizer inputs. In China, this modern approach is helping to balance soil fertility, overcome regional differences in biochemical parameters, improve fertilizer utilization rates and crop productivity, obtain higher crop yields and economic gains (Guo *et al.,* 2016; Bierman *et al.,* 2012), and promote the sustainable mechanization and automation of modern agricultural production.

The concept of **"precision agriculture"** is known to the world since 1980's when American agricultural producers in the 1980s started using variable-rate fertilization technology (Yu *et al.,* 2010; Zhao *et al.,* 2009). The core control technology is largely based on prescriptive information and real-time parameters (Yuan *et al.,* 2014; Gu *et al.,* 2011). The variable-rate fertilization control method based on prescriptive information (because of the limitations of the complicated process, difficult operation and high economic costs) cannot apply in scientific and reasonable variable fertilization according to the real-time nutrient demands during crop growth cycles (Li *et al.,* 2017). However, **variable-rate fertilization control technology** (VRT) based on real-time parameters avoids the above-mentioned intricacy process, can be used to address problems associated with uneven growth and unbalanced soil nutrients. Recent developments such as real-time sensor equipment, soil nutrient parameters and crop growth information that can now be monitored online to generate target fertilizer applications in real-time (Zaman *et al.,* 2011), and scientifically guide the operation equipment to achieve on-demand, variable-rate fertilization. The critical factor in the success of this modern approach (VRT) is the acquisition of stable, representative, and accurate field information including crop phenotypic information (Tang *et al.,* 2019). Studies have shown that there is a general relationship between real-time fertilizer requirements of crops and the phenotypic nutritional status of their corresponding growth stages, and that the real-time growth is closely related to the biological elements characteristics of leaves (Raun *et al.,* 2005). Crucially, the biological characteristics of crops show significant correlations with canopy spectral reflectance information (Chen *et al.,* 2011). Thus, the reflectivity and spectral parameters of different wavelengths have different quantitative relationships with leaf nutrient elements (*e.g.,* nitrogen, phosphorus, and potassium content). The chlorophyll index (SPAD), leaf area index (LAI), ratio vegetation index (RVI), normalized difference vegetation index (NDVI), and other indexes can predict crop phenotypic parameters fairly well (Xue *et al.,* 2011; Maleki *et al.,* 2007; Zhu *et al.,* 2010) including reflecting real-time growth information, that can be used to guide scientific and reasonable topdressing operations. Therefore, the appropriate fertilizer applications required for crops can be estimated by the effective models of statistical analysis based on phenotypic biological parameters (Ma *et al.,* 2015; Xue *et al.,* 2008).

The advancement in computer science and remote sensing (RS) technologies helped to evolve new ideas and methods for real-time, non-destructive nutrient detection during crop growth have been developed (Tang, 2018). It is quite interesting to note that the high resolution imageries and spectral signatures can be used as the base data to determine the health status of the plants non-destructively with a short span of time. This allows the scientists to take quick decision to correct the maladies in crops. Thus, portable high-throughput phenotypic monitoring platforms combining agricultural RS technology and crop growth bio-indexes have emerged (Chen *et al.,* 2016). The real-time sensors available at present can assist to detect the field problems quickly, accurately, and non-destructively (Haboudane *et al.,* 2002; Lan *et al.,* 2019). At present, many well-developed products are widely applied in real-time variable-rate spreading operations (Basso *et al.,* 2011; Song *et al.,* 2009), including MSR-16 radiometers produced by American Cropscan Company, Crop Circle ACS-470 crop canopy sensors from the Holland Scientific Company in the United States, and GreenSeeker spectral sensors jointly developed by the American Trimble Company and Oklahoma State University. Considerable exploratory research has also been carried out in China (Yang *et al.,* 2013).

The application of RS technology has gone beyond plant health monitoring it is widely used for land cover and crop biomass measurements besides biological and physicochemical parameters. The theoretical basis of RS is the targeted detection of spectral characteristics differences of electromagnetic waves based on crop phenotypes (Wu *et al.,* 2013; Maes and Steppe, 2019). Research has found that when exposed to electromagnetic waves, the spectral reflectance, absorption, transmission, and emission characteristics responding in different wavebands differ between crops (Lie *et al.,* 2009). Especially in the visible-near-infrared spectrum bands, reflectivity is mainly affected by phenotypic pigments, cell structure, water content as well as other biological parameters. There are strong absorption characteristics in the visible red light wavebands and strong reflection characteristics in the near-infrared wave bands, which are often used for monitoring crop growth, production yield and quality, and pests and diseases (Zhang *et al.,* 2010). Generally, spectral reflectivity is the ratio of the reflected energy of an object to the incident energy at a particular wavelength interval. When the solar spectrum with a wavelength of λ is projected into the crop canopy in parallel, Lambertian reflectance occurs (Ni *et al.,* 2013), thus the reflection model of crop phenotype conforms to Lambert's cosine law. The law states that the radiant intensity or luminous intensity observed from an ideal diffusely reflecting surface or ideal diffuse radiator is directly proportional to the cosine of the angle θ between the direction of the incident light and the surface normal. The law is also known as the cosine emission law or Lambert's emission law. It is named after Johann Heinrich Lambert.

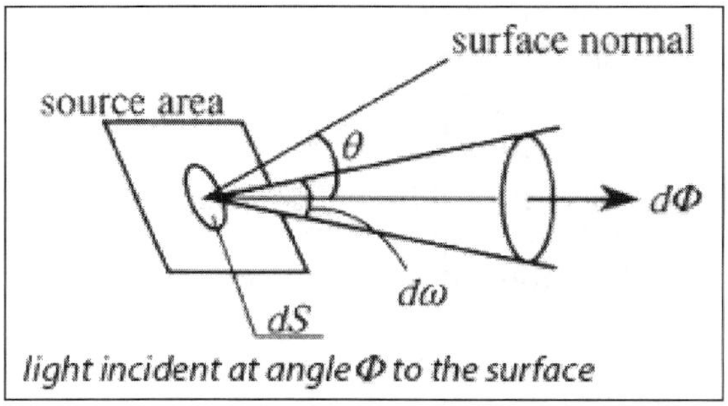

Fig. 5.1

Real-time variable-rate fertilization technology applies these radiation transmission mechanisms to obtain biological information about crop phenotype, nutrient use, and production yield quality. This approach depends on accurate determination models for different nutrient and biological elements inversed by RS technology, such as the nitrogen/chlorophyll concentration, nitrogen/chlorophyll accumulation, leaf area index, biomass, and canopy density (Hu, 2018). Furthermore, along with expert decision systems built into variable-rate fertilization equipment, scientific and rational nutrient regulation and fertilizer management can be enabled. This effectively solves the problems associated with regional differences of biochemical parameters of field crops, improves fertilizer utilization efficiency and crop production capacity, and can support precision agriculture modernization and sustainable development.

5.2. Crop Phenotypic Information and Nutritional Status

Remote sensing technology is widely applied in crop phenotypic monitoring and the diagnosis of plant nutritional status in large-scale agricultural production. This includes growth detection and yield prediction of wheat, corn, rice, cotton, and other staple crops (Liu *et al.,* 2012; He *et al.,* 2017). The monitoring of canopy biological characteristics can be inverted for the assessment of growth and plant nutritional status (Li *et al.,* 2011; Fu *et al.,* 2014). This relies on the internal relationships between crop features (*e.g.*, canopy geometry, leaf biochemical composition, and internal tissue components) and canopy spectral reflectance characteristics (Yang *et al.,* 2011; Ni *et al.,* 2013). Data are obtained through spectral data mining (*e.g.*, statistical discriminant, correlation analysis, and regression modeling) to establish corresponding algorithm and metrics, such as the crop nitrogen nutrition index (NNI), ratio vegetation index (RVI), and leaf nitrogen accumulation (LNA) (Li *et al.,* 2013).

Currently, the commonly used **inversion methods** are mainly divided into two categories. **First,** the empirical statistical method is based on the relationships between crop phenotype biological parameters and spectral reflectivity or other spectral characteristics in sensitivity bands. This is a relatively simple approach; however, established empirical relationships rely excessively on specific conditions, such as the measurement instrument, measurement time, crop variety, and environmental environment, meaning that applicability can be limited (Liang *et al.,* 2012). **Second,** the radiation transmission model inversion method is based on RS monitoring process, commonly used numerical optimization algorithms, the partial least-squares method, principal component analysis (PCA), support vector machine training sets, and the artificial neural networks. These techniques are applied to simplify the derivation process and improve inversion efficiency, which has a stronger universality (Yebra *et al.,* 2009; Nguy-Robertson *et al.,* 2011). The application of these methods in research on high throughput biological monitoring and nutrient status discrimination of crop phenotypes has included multiple scales, from crop populations to plant tissues and organs, and has yielded rich and diverse results.

Crop canopy cover is a phenotypic parameter that quantitatively describes the dynamic differences in growth and development between genotypes. It is an important indicator to characterize and evaluate crop nutritional status (such as nitrogen status, early vitality, senescence state, and surface biomass) and has important applications in monitoring crop growth. Currently, there are many indexes that are closely related to crop canopy cover, of which the NDVI and the LAI are the most well-developed (Li *et al.*, 2008; Tang *et al.*, 2011). The NDVI can be monitored in real-time and has been applied as an objective means of (indirectly) assessing variations in the characteristics of wheat under stress and seasonal nitrogen demand. It can also be used to estimate crop nitrogen content, surface nitrogen distribution, and crop nitrogen utilization efficiency. The close correlation between NDVI and crop physiological characteristics can also be used to explain the diversity impact factors on crop production, such as changes in moisture and nitrogen regimes during different growth stages. Li *et al.* (2010) discussed the relationship between the NDVI and crop LAI using Thematic Mapper (TM) imagery and directly used the inversed LAI as the growth grading standard to monitor the growth of winter wheat in Xinghua City, Jiangsu Province, China. Their study also assessed indexes of crop nitrogen content and nitrogen demand application to guide scientific and rational variable application of nitrogen fertilizer on demand, this improves the level of fertilizer utilization efficiency and produces significant economic benefits.

5.3. Nitrogen Nutritional Status

Nitrogen is referred as "kingpin" aong essential nutrients required for plants to complete its lifecycle. Nitrogen is one of the most critical nutrients deciding the fate of agricultural productivity. Most of the tropical soils including Indian soils are widely deficient and there is a quick and spontaneous response for N fertilization. Farmers are fully convinced with N fertilization as it gives visible and obvious difference between before and after N fertilization. As a result, in variable rate fertilization, nitrogen is given utmost importance. The nitrogen demand application is the central in the monitoring and diagnosing of crop growth (and the important mainstay for making management decisions of field production), and provide the necessary technical basis for achieving the modernization and sustainable development of precision agriculture. Shibayama *et al.*, 1986 have showed that the nitrogen content per unit area of crop leaves had strong regression relationships with reflectivity in spectral bands R400, R620, R760, and R880, which were not affected by crop type or variety. Liu *et al.*, 2002 used the characteristics of the reflection peak at 1690 nm to invert the carbon–nitrogen ratio of wheat canopies at each growth stage. This enabled a segmented RS inversion model of leaf and stem-sheath to be established for the rise and milk stages, and the accuracy of estimating the nitrogen content in fresh leaves using spectral analysis method was improved to > 85 per cent. Cho *et al.*, 2006 proposed a linear extrapolation method to calculate red edge position information about crop canopy reflectance spectra, and established models for predicting the nitrogen concentrations of wheat plants, maize leaves, and mixed pasture plants. Chen *et al.*, 2010 showed that the dual-peak canopy nitrogen index (DCNI) eliminates the interference of leaf area changes with respect to the diagnosis of nitrogen concentrations and, thus, enables the accurate detection of plant nitrogen concentrations. Furthermore, Chen *et al.,* 2010 were able to derive the nitrogen nutrition index (NNI) of winter wheat and maize using RS techniques, which could well determine the nitrogen nutrition status.

With the advancement of research on crop phenotypes and nitrogen nutrition, increasing attention has been paid to the heterogeneity of vertical distribution of nitrogen in crop canopies, nitrogen content at different height of the crop canopy plays an important implications for scientific guidance of variable-rate fertilization and field management (Chunhua *et al.,* 2008; Li *et al.,* 2013). For example, Wang *et al.,* 2007 used a partial least-squares (PLS) algorithm to invert the nitrogen concentration of leaves at different canopy levels based on the vertically observed canopy spectra of winter wheat. Wang *et al.,* 2004 also estimated the nitrogen density of different leaf layers using the difference vegetation index (DVI) of winter wheat combined with the interception of luminous energy at each canopy level. Zhao *et al.,* 2013 achieved higher accuracy estimates by constructing an inversion index of chlorophyll concentrations in the upper, middle, and lower leaf layers of winter wheat based on multi-angle canopy reflectance spectra. Based on field experiments and RS, Ma *et al.,* 2007 further analyzed the correlation between nitrogen concentrations and the spectral and fluorescence characteristics of leaves at different vertical heights, which provided a new idea for exploring the RS method to determine the vertical distribution of crop canopy nitrogen.

Using field-based equipment, observation towers, and hand-held sensors as carriers (according to the acquired phenotypic biological information and inversed nutrient evaluation models), rational on-demand nitrogen application and regulation management can be achieved. Such equipment includes radiometers (*e.g.,* the MSR-16), nitrogen sensors (*e.g.,* the Yara N-sensor), crop canopy sensors (*e.g.,* the Crop Circle ACS-470), spectral sensors (*e.g.,* GreenSeekers), and non-imaging spectrometers (*e.g.,* the ASD Fieldspc FR2500) (Wu *et al.,* 2010; Qaio *et al.,* 2008; Wu *et al.,* 2010). Specific examples include Kitchen *et al.,* 2010, who applied the green channel NDVI (GNDVI) obtained using the 'Crop Circle' device to monitor crop nitrogen deficiency, and then guided the variable-rate fertilization. Their results showed that the method could be used to reduce the nitrogen application and improve the crop yield. Peng *et al.,* 1996 also proposed a means of real-time nitrogen management (RTNM) based on measured crop SPAD values to rapidly determine nitrogen nutritional status. Scarf *et al.,* 2002 estimated the nitrogen nutritional status of maize in an unfertilized area based on the difference in corn greenness values between a saturated and control fertilization area, and used this as a basis for variable-rate fertilization management. Lukina *et al.,* 2001 used canopy NDVI to invert the nitrogen accumulation of winter wheat to estimate potential yield under variable-rate fertilization management. Their research was developed further by Chen *et al.,* 2003, who used field object spectral data to improve the universality of the nitrogen application optimization algorithm. Aning *et al.,* 2007 also established a variety of variable-rate fertilization algorithms for winter wheat based on the optimized soil-adjusted vegetation index (OSAVI) and chlorophyll (SPAD) values. Their results showed that the economic and ecological benefits of variable-rate fertilization based on the

5.4. Real Time Variable Fertilization Systems Based on Crop Phenotypic Information

Based on the continuous development of computer science and RS technologies, and the agronomic requirements of different types of crops, detection platforms for the acquisition of biological information from crop canopies have been increasingly applied in real-time variable-rate fertilization (Hu *et al.,* 2018). According to canopy biological parameters obtained using real-time monitoring systems, expert decision-making systems are then used to determine the target fertilizer application online. Once determined, the core controller adjusts the actuator, which controls the actual application of fertilizer to the crop. Relative to traditional approaches to fertilizer application, such set-ups improve fertilizer utilization efficiency, increase crop yields and economic gains, and improve ecological standards. In accordance with the different application carriers, phenotypic monitoring platforms are mainly divided into two types: (1) aviation based and (2) ground-based approaches. Common aviation-based phenotypes monitoring technologies are divided into satellite-borne and airborne types, and ground-based technology can be divided into hand-held and vehicle-mounted types (Guo *et al.,* 2018).

Multi-source satellite images acquired using the satellite- borne platforms can be used to extract crop phenotypic parameters; however, due to the limited time and spatial resolutions of data acquisition, it is difficult to apply over small-area test regions and for high frequency dynamic monitoring (Sankaran *et al.,* 2018). In comparison, airborne phenotypic monitoring platforms (*e.g.,* helicopter or unmanned aerial vehicle [UAV]) carrying multiple sensors simultaneously, can acquire multi-source image information for entire test sites in a relatively short period of time (Boegh *et al.,* 2002; Deery *et al.,* 2016; Rajan *et al.,* 2009). This approach has been applied to measure the canopy temperatures and lodging situations in thousands of field plots as well as canopy structure at various test sites (Chapman *et al.,* 2014; Jonckheere *et al.,* 2004). UAVs or drones are now used to carry multispectral cameras (such as the Mini-MCA6), hyperspectral cameras (such as the Cubert UHD185) and active non-imaging sensors (such as the RapidSCAN) that enables the derivation of the LAI, aboveground biomass, and nitrogen nutrient status for crops such as wheat, rice, cotton, and corn (Krienke *et al.,* 2017; Liu *et al.,* 2018; Zheng *et al.,* 2018) For example, in Japan, TM-based RS data has been used to monitor nitrogen, amylose, amylopectin, and other quality indicators in rice, which has been used to guide the application of nitrogen fertilizer (Konishi *et al.,* 2000). This reduced the nitrogen content of rice grains in a larger experimental area from 7.7 to 7.3 per cent and thereby improved crop quality economic gains. However, airborne platforms involve higher costs, require more technical maintenance effort, and the RS information obtained by individual sensors is often limited. Multi-sensor coordination and data fusion are, therefore, required. Moreover, canopy disturbance effects become problematic when acquiring high-resolution images at low altitudes (Araus *et al.,* 2014; Duan *et al.,* 2016; Hu *et al.,* 2018). Ground-based phenotypic platforms usually involve common agricultural equipment adapted with a series of phenotypic sensors and a GPS for positioning and navigation (Busemeyer *et al.,* 2013; Sharma *et al.,* 2016), these platforms can simultaneously acquire multi-source information (such as multispectral, hyperspectral, and thermal infrared images). As different sensor groups can vary greatly at different test points and under variable environmental conditions, therefore, means of calibrating sensor groups are required to ensure reliability and comparability, which is the main technical challenge when using this vehicle-based phenotype platform (Tian *et al.,* 2016).

At present, data acquisition technology providing phenotypic information for field crops (widely applied in real-time variable-rate fertilization for precision agriculture) is still dominated by ground-based and vehicle- mounted operations. Vehicle-mounted phenotypic and omics platforms are simple to operate and can obtain high-resolution and continuous spectral data from closeup observations. Sensors are mounted at different locations and heights of the platform depending on the crop type and growth period (Rebetzke *et al.,* 2016). Related accessories include power systems, data acquisition terminals, GPS receivers, and encoders to ensure the running work of the vehicle. A high-performance terminal that can receive various sensor data is also needed. When equipped with a Beidou satellite, radar, GPS, gyroscope, and other sensors, vehicle- mounted platforms can achieve automatic navigation, generate distribution maps of crop phenotypic features, and coordinate with the optimized expert decision system to conduct reasonable field nutrient management (Sharma *et al.,* 2015; Jiang *et al.,* 2019).

Currently, intelligent agricultural equipment for variable-rate fertilizer application systems is relatively well-developed, and a wide range of technical equipment has been commercialized and applied internationally and at a large-scale **(Figure 5.2).** However, airborne platforms involve higher costs, require more technical maintenance effort, and the RS information obtained by individual sensors is often limited.

Ehlert *et al.,* 2009 developed a variable-rate nitrogen fertilizer spreader for winter wheat based on a crop density detector with a mechanical pendulum. In their system, an onboard computer continuously adjusts fertilizer application based on crop density parameters measured at a fixed point. Multiple field trials using calcium ammonium nitrate fertilizer (nitrogen fertilizer) under different seasons showed that this system could reduce fertilizer application by 10 12 per cent and improve crop yield and quality. To improve the utilization rate of nitrogen fertilizer during the cotton production process in Tennessee, USA, and to optimize the impact of soil reflection using the GreenSeeker sensor, Benitez *et al.,* 2010 designed an optimization algorithm and an ultrasonic device to control spraying distance form cotton canopy, as shown in **Figure 2.** A variable-rate fertilization machine based on spectral detection technology was developed for maize by Lee and Searcy (Lee *et*

Figure 5.2: (a) Multi-level crop phenotype monitoring platform; (b) UAV crop phenotype monitoring platform; (c) IoT phenotype monitoring platform; (d) Mobile vehicle-mounted monitoring platform (Source: Shi *et al.,* 2020).

al., 2000) In their set-up, a spectrum monitoring system including a halogen lamp and an N sensor was installed behind a tractor to monitor the nitrogen nutrition status of corn in real-time **(Figure 5.3).** GreenSeeker sensors and Multiplex sensors have also been used to monitor the nutritional status of grapes in real-time and to guide field fertilizer management. In this case, the system was able to obtain various field indices including the chlorophyll fluorescence index (SFR), nitrogen balance index (NBI), and flavonoid index (FLAV).

In China, research on precision agricultural production and variable-rate fertilization technology has been relatively slow to progress and remains in its infancy. Recent developments have mainly resulted from the introduction of foreign advanced agricultural machinery and their supporting control systems, although there remains a lack of systematic theoretical research and related experiments in this field (Yuan *et al.,* 2013). In recent years, with the continuous development and improvement of relevant national policies including the 'Application Reduction

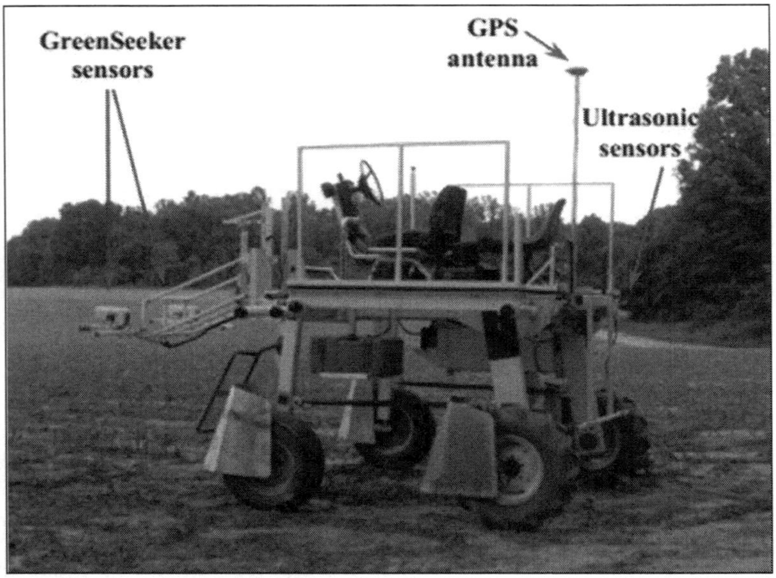

Figure 5.3: Variable Fertilizer Applicator for Cotton.

and Efficiency Increase' policy, independent and innovative research in intelligent agricultural machinery and control systems has begun to progress rapidly with groping and exploring of precision agricultural in succession. Examples include Qiao *et al.,* 2008 who used an MSR16R multispectral instrument to obtain canopy spectral reflectance information for soybean crops. They established a nutrition diagnosis model to estimate the required levels of nitrogen for growth and an expert decision-making system for stable, variable-rate fertilization based on fuzzy control theory (Figure 5.3). Zhang, 2012 developed an online measurement variable-rate fertilizer applicator based on crop growth (Figure 5.4). They used non-destructive, photoelectric inspection technology to obtain real-time NDVI data and a core processor controlled different fertilizer applications based on a fuzzy control algorithm. Based on the correlation between crop canopy spectral reflectance and growth, their system achieved the real-time online 'living diagnosis' of community growth trends and the LAI across the whole crop growth period. Their system was able to guide reduced fertilizer application in areas of stronger crop growth, and more fertilization in areas with poor growth, thus enabling precise variable-rate fertilization according to crop growth.

Yinyan *et al.,* 2018 evaluated chemical (nitrogen) fertilizer demand based on NDVI values of winter wheat using real-time monitoring and implemented an intelligent control program for variable-rate fertilization. At the same time, their decision-making tool provided real-time targeted top-dressing applications, realized the online adjustment of the fertilization amount (Wang *et al.,* 2018). Furthermore, Yinyan *et al.,* 2018 developed a centrifugal and high-efficiency

a Back view of the sprayer. b Closer view of the lighting chamber.

c Inside of the lighting chamber: front view. d Inside of the lighting chamber: side view.

Figure 5.4: Variable-rate Fertilizer Applicator for Corn.

variable-rate fertilizer spreader based on real-time growth information for rice. In this case, a GreenSeeker spectral detection system obtained real-time DNVI values of rice canopies and was coupled with the previously established expert decision making system to generate target fertilizer requirement application, then real-time spreading application was adjusted online (Shi *et al.,* 2018) to realize accurate variable-rate fertilizer spreading for rice. It can be seen that real-time variable-rate fertilization technology based on the vehicle-mounted phenotype omics platform has great application potential in the research of crop phenotype bioinformation monitoring and variable-rate fertilization technology research of precision agricultural. Moreover, of course, other vehicles that can be mounted with sensors and high-throughput phenotypic acquisition tools include airships, parachutes, hydrogen balloons, and ground observation towers, but these are limited by their inflexibility (such as line spacing, plant spacing, and crop height limitations) and cost, which makes them difficult to apply across different crops and at different growth stages. Furthermore, the image resolution, safety performance, and load capacity of these other approaches all need to be improved further (Maleki *et al.*, 2007).

5.5. Summary of the Literature Review

The literature review suggest that the phenotypic information monitoring, precision agriculture, and variable-rate fertilization technologies has progressed well in China. However, real-time variable-rate fertilizer systems and its application based on crop phenotype bioinformation are still at the experimental stage being currently limited from the following key challenges:

1. Research on variable-rate fertilization technology based on crop phenotypic biological information mainly focuses on specific field conditions and crop varieties. However, given that China has a relatively developed agricultural economy and a substantial cropping area, research into the application of precision agriculture that accounts for regional and local features is lacking, especially for multi-species crops in agricultural underdeveloped areas. For example, most of the existing research targets wheat, corn, rice, barley, and other cereal crops while very few studies have been reported for beans, potatoes, and other important commercial crops.
2. Obtained biological parameters of crop phenotypes are single. Each imaging technique has its own limitations that are affected by temperature, illumination, and other complex field conditions. Some RS techniques are limited by their measurement accuracy, and spatial and spectral resolution is yielding large measurement errors, and they cannot be widely applied. It is, therefore, necessary to explore the fusion of different phenotypic data sources, at different stages of data processing, to overcome or reduce detection error associated with single-source approaches. This is essential to improve the accuracy of crop phenotypic parameters and the stability of variable-rate fertilization equipment.
3. Rapid processing of the enormous RS data faces huge challenges. Hyperspectral imagers and Lidar sensors rapidly acquire large amounts of high-temporal resolution information. The processing of these datasets is, therefore, limited by computer hardware performance and processing power. It remains difficult, for example, to achieve rapid online processing of 'big data' in real-time, and the mixed spectra of crops at different growth stages along with variable soil compositions adds significant complexity. Indeed, the mixed spectra of field components are difficult to decompose, resulting in lower accuracy targeting of fertilizer requirements for inversion calculations.

4. Inversion models of fertilizer requirement are lacking in common methods. For practical application, when parsing phenotypic information for different crops and under different growth stages, modeled needs to be performed separately, which ignores the synergistic influence of multiple physiological parameters? The applicability of models is also limited by environmental conditions and variability and uncertainties, such as the driving variables or parameters of RS inversion models, and uncertainty associated with error propagation. This leads to crop growth models and nutrient recommendation systems that cannot be closely matched with variable-rate fertilization machinery and, hence, the cycle of expert-decision analysis in the system slows down.
5. Currently, intelligent control systems have little supporting research and limited development capabilities. Many sensors remain expensive, vehicle mounted hyperspectral sensors, laser radar sensors, and chlorophyll fluorescence sensors must all currently be imported to China, resulting in high costs. This also means that the core technology of these sensors cannot be easily adapted, optimized, and upgraded, limiting the applicability under specific operational demands of the Chinese agricultural sector. Detection sensors for crop phenotypic monitoring with independent intellectual property rights are also lacking, which restricts the extensive promotion and application of variable-rate fertilization equipment based on crop phenotypic information in modern planting areas in China.

Whilst still at the exploration stage, as an emerging trend in precision agriculture, real-time, variable-rate fertilization technology based on crop phenotypic information will continue to develop in China into the future. This will serve as the base data for other countries to adopt. Country like India wishes to introduce precision agriculture and the literature review help us to integrate the lessons learnt from China. Further research and promotion are required, especially for high-throughput phenotypic data acquisition and the integration and application of real-time variable systems. Further technical development and research collaborations are, therefore, needed. Building on the advanced experience of other countries and based on the specific machinery and agronomic requirements of Chinese agriculture, matching variable-rate fertilizer equipment with independent innovation should be targeted. This will reduce production costs and demonstrate and applicability in modern agricultural areas, thereby fostering the effective and efficient utilization of fertilizer resources to achieve healthy and sustainable development. In the future, we believe that real-time variable-rate fertilization techniques based on crop phenotypic information will continue to receive attention under the key themes outlined in the following sections.

5.6. Application of Drones and Sensors in Real-Time Variable Rate Fertilization Technology

The application of drones and sensors for the on-site detection and decision making for fertilizer prescriptions have been studied for a wide array of crops and the data suggest the following:

1. Drone application in on-site detection of nutrients and fertilizer prescriptions have been studied in rice, wheat, maize, cotton, sunflower, soybean, potato and turf grass
2. Sensors used for the detection of nutrient status particularly nitrogen include hyperspectral, multi-spectral and RGB, CIR. These sensors have shown close correlation between the spectral signatures and nitrogen content of the leaves.
3. Suitable UAVs for nutrient monitoring are fixed wing or rotary type. In many cases it is supported ground truth using sensors or SPAD meters to establish relationship between measurements
4. Standard indices used for nutrient status monitoring are NDVI, NGDRI, VI and RDVI and these were used regardless of types crops being cultivated.
5. Nitrogen estimation was performed using a classic micro-kjeldahl method or non-destructive SPAD meter method. These methods are quite robust and commonly used as a reference to validate the spectral reflectance.
6. Statistical models that were used to establish relationship between spectral signature and ground truth include linear regression, polynomial regression, multi-linear regression *etc.*

Overall, the literature review clearly demonstrates that the nutrient status of the plants particularly N can be determined non-destructively using sensors mounted on to the drones. Despite close and significant relationships have been established for several crops, use of these techniques in smart farming is very restricted due to practical difficulties.

5.7. Case Studies with Reference to Specific Crops

5.7.1. Rice

With the rise of large-scale crop plantation in China, inexpensive but efficient remote sensing measures for predicting the nitrogen status of crops are needed for optimal fertilizer management. Li *et al.* (2015) study the exploited a conventional digital camera with a charge coupled device was integrated into an unmanned aerial vehicle (UAV) to capture digital aerial images of paddy rice (*Oryza sativa* L.) at an altitude of 50 m. The fluorescence emissions of the rice leaves under light excitation were used by Multiplex® to non-destructively assess the chlorophyll and polyphenol content. The nitrogen balance index (NBI) of the rice leaves, known as the ratio of chlorophyll to polyphenols, was used to accurately determine canopy

Table 1: Use of Drones and Sensors to Fertilization Schedule in Agricultural and Horticultural Cropping Systems

Sl.No.	*Crops*	*Sensor*	*UAV Type*	*Model Input*	*Model Output*	*Model Type*	*Reference*
1	Rice	RGB	Rotary	DGCI	Nitrogen concentration	Linear regression	Li *et al.*, 2015
2	Rice	Multispectral	Fixed wing	Vegetation indices	SPAD (chlorophyll content)	Linear regression	Saberioon *et al.*, 2017
3	Rice	Multispectral	Rotary	Vegetation indices	N treatment	Linear regression	Swain *et al.*, 2007
4	Wheat	Hyperspectral	Rotary	Selected bands	Nitrogen concentration	Multilinear regression	Zhu *et al.*, 2009
5	Wheat	Hyperspectral	Rotary	Selected bands	Nitrogen concentration	Multilinear regression	Liu *et al.*, 2017
6	Wheat	Multispectral	Rotary	RDVI	Nitrogen concentration	Several ML models	Zheng *et al.*, 2018
7	Wheat	Multispectral	Parafoil and fixed wing	Vegetation indices	Nitrogen uptake	Exponential regression	Lelong *et al.*, 2008
8	Wheat	Multispectral	Rotary	NDVI	Nitrogen concentration and uptake	Linear regression	Benincasa *et al.*, 2018
9	Wheat	Multispectral	Rotary	NDVI, REIP	Nitrogen concentration	Linear regression	Geipel *et al.*, 2016
10	Maize	RGB	Fixed wing	NGRDI	Nitrogen status (chlorophyll content)	LS regression	Hunt *et al.*, 2005
11	Maize	RGB	Rotary	Pixels	Nitrogen deficiency level	Logistic regression	Zermas *et al.*, 2015
12	Maize	Multispectral, hyperspectral	Rotary	Variety of indices	Nitrogen concentration	Polynomial regression	Gabriel *et al.*, 2017
13	Maize	CIR	Rotary	Several vegetation indices	Nitrogen concentration and uptake	PLS regression	Corti *et al.*, 2018
14	Cotton	Multispectral	Rotary	Several VIs	Nitrogen concentration and uptake	Linear regression	Ballester *et al.*, 2017
15	Sunflower	Multispectral	Rotary	NDVI	Nitrogen treatment	Linear regression	Agüera *et al.*, 2011
16	Soybean	Multispectral, thermal, RGB	Rotary	Spectral indices and features	Nitrogen concentration	PLSR, SVR, ELR	Maimaitijiang *et al.*, 2017

Sl.No.	*Crops*	*Sensor*	*UAV Type*	*Model Input*	*Model Output*	*Model Type*	*Reference*
17	Sugar beet	Multispectral	Rotary	VIs, green pixel fraction	Nitrogen concentration	Multilinear regression	Jay *et al.*, 2018
18	Potato	Multispectral	Rotary	NDVI, GNDVI	N status (chlorophyll content, LAI)	Linear regression	Hunt *et al.*, 2016 Hunt *et al.*, 2018
19	Turfgrass	Multispectral	Rotary	NDVI	Nitrogen content	Linear regression	Caturegli *et al.*, 2016
20	Grass	Hyperspectral	Rotary	Average reflectance spectra	Sodium and potassium content	PLS regression	Capolupo *et al.*, 2015

nitrogen concentrations. The dark green colour index (DGCI) available from the aerial images was used to assess the nitrogen concentrations in the field. It was found that DGCI values predicted the nitrogen concentrations and NBI with R2 (coefficient of determination) = 0.672 ($p < 0.001$) and R2 = 0.711 ($p < 0.001$), respectively. The results indicated that aerial photography from UAV had the ability to assess plant nitrogen status in the field in large-scale crop plantations.

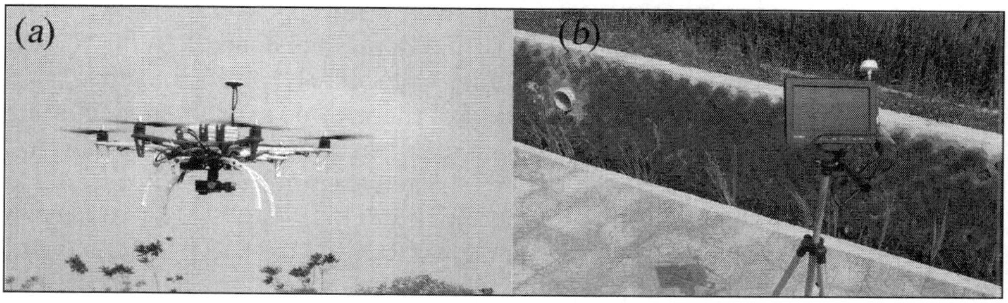

Fig. 5.5: (a) The six-rotor UAV with a domestic digital camera; (b) the ground monitor displaying real-time imagery and other data transmitted from the UAV.

The DGCI values were calculated the following steps are followed. To simplify the interpretation of the digital colour data, Karcher and Richardson (2003) suggested converting the R, G and B values to a more intuitive hue, saturation and brightness (HSB) colour spectrum, which is based on the human perception of colour. A dark green colour index (DGCI), derived from values of HSB, was used to assess the greenness of the aerial images. The DGCI was used to assess the nitrogen concentration of the canopy, and it was calculated as:

$$\mathrm{DGCI} = \frac{[\mathrm{H} - 60/60 + (1 - \mathrm{S}) + (1 - \mathrm{B})]}{3}$$

Where H, S and B are the HSB of a digital image (converting R, G and B to H, S and B can be found in Karcher and Richardson's report (2003)).

Both fluorescence-based FLAV or NBI_R and LARS-based DGCI have the advantage of being able to assess the plant nitrogen status at the canopy and field levels, respectively. The LARS platform can be a promising and inexpensive substitute for satellite imaging and can be applied in the diagnosis of the plant nitrogen status at a large scale in the field. Both FLAV and NBI_R predicted the canopy nitrogen status with high precision (R2 = 0.792, $p < 0.001$; R2 = 0.691, $p < 0.001$, respectively), but SFR_R should be used carefully. DGCI values predicted the nitrogen concentrations and NBI_R with satisfactory precision (R2 = 0.672, $p < 0.001$; R2 = 0.711, $p < 0.001$, respectively). The results suggest that aerial photography by UAVs have great potential in supporting crop decision-making processes, but more in-depth research concerning the effect of illumination conditions in different growth stages needs to be conducted.

In another study, **Zheng (2018)** evaluated three sensors (RGB, color-infrared (CIR) and multispectral (MS) cameras) onboard UAS for the estimation of N status at individual stages and their combination with the field data collected from a two-year rice experiment. An Oktokopter UAS was used to acquire aerial photography at early growth stages (from tillering to booting) and field samplings were taken at a near date. Two color indices (normalized excess green index (NExG), and normalized green red difference index (NGRDI)), two near infrared vegetation indices (green normalized difference vegetation index (GNDVI), and enhanced NDVI (ENDVI)) and two red edge vegetation indices (red edge chlorophyll index (CIred edge), and DATT) were used to evaluate the capability of these three sensors in estimating leaf nitrogen accumulation (LNA) and plant nitrogen accumulation (PNA) in rice. The results demonstrated that the red edge vegetation indices derived from MS images produced the highest estimation accuracy for LNA (R^2: 0.79–0.81, root mean squared error (RMSE): 1.43–1.45 g m^{-2} and PNA (R^2: 0.81–0.84, RMSE: 2.27–2.38 g m^{-2}). The GNDVI from CIR images yielded a moderate estimation accuracy with an all-stage model. Color indices from RGB images exhibited satisfactory performance for the pooled dataset of the tillering and jointing stages. Compared with the counterpart indices from the RGB and CIR images, the indices from the MS images performed better in most cases. These results may set strong foundations for the development of UAS-based rice growth monitoring systems, providing useful information for the real-time decision making on crop N management.

5.7.2. Maize

Maize is extensively cultivated in India and it is one of the major consumer of fertilizers particularly nitrogen. There are lots of private hybrids have come into the market that are highly responsive to N fertilization. In the past few years, the fertilizer N dose of maize has increased from 150 kg/ha to now 250 kg/ha and some places farmers go up 400 kg N/ha. Thus, there is an urgent need developing a sensor or drone-enabled remote sensing approach to detect N precisely so as to rationalize the fertilizer application. Remote sensing is a key technology for precision agriculture to assess actual crop conditions. Commercial, high-spatial-resolution imagery from aircraft and satellites are expensive so the costs may outweigh the benefits of the information.

Hunt (2005) evaluated these very-low-cost, very high-resolution digital photography for use in estimating nutrient status of corn and crop biomass of corn, alfalfa, and soybeans. He optimized an aerobatic model aircraft for acquiring pictures using a consumer-oriented digital camera. Colored tarpaulins were used to calibrate the images; there were large differences in digital number (DN) for the same reflectance because of differences in the exposure settings selected by the digital camera. To account for differences in exposure a Normalized Green–Red Difference Index [(NGRDI = (Green DN) Red DN)/(Green DN + Red DN)] was used; this index was linearly related to the normalized difference of the green and red reflectance, respectively. For soybeans, alfalfa and corn, dry biomass from zero to 120 g m^2 was linearly correlated to NGRDI, but for biomass greater than 150 g m^2

in corn and soybean, NGRDI did not increase further. In a fertilization experiment with corn, NGRDI did not show differences in nitrogen status, even though areas of low nitrogen status were clearly visible on late-season digital photographs. Simulations from the SAIL (Scattering of Arbitrarily Inclined Leaves) canopy radiative transfer model verified that NGRDI would be sensitive to biomass before canopy closure and that variations in leaf chlorophyll concentration would not be detectable. There are many advantages of model aircraft platforms for precision agriculture; currently, the imagery is best visually interpreted. Automated analysis of within-field variability requires more work on sensors that can be used with model aircraft platforms.

This study was a comprehensive examination on the use of radio-controlled model aircraft and digital cameras as a very-low-cost, very-high-spatial-resolution alternative for remote sensing in precision agriculture. The results were mixed, because there were limitations imposed by the use of fixed-wing model aircraft and there were limitations imposed by the use of a consumer-oriented digital camera. The maximum amount of information would be obtained from visual interpretation of the images, particularly regarding crop nitrogen status. Automated digital image processing based on indices usually removes some of the subjectivity involved with interpretation, and hence is an important step in exacting useable information from the imagery. On the whole, the use of model aircraft and digital cameras overcame many of the problems associated with commercial satellite and airborne imagery, thus these technologies warrant continued investigation and improvement. Based on the development of sophisticated, lightweight sensors used in unmanned airborne vehicles (UAVs) for military operations, major advances using model aircraft or other alternative platforms for remote sensing in precision agriculture should be expected.

5.7.3. Wheat

Lu *et al.* (2019) employed a UAV-based five-band camera to acquire multispectral images at seven view zenith angles (VZAs) (0°, ±20°, ±40° and ±60°) for three critical growth stages of winter wheat. They used four representative vegetation indices encompassing the Visible Atmospherically Resistant Index (VARI), Red edge Chlorophyll Index (CIred-edge), Green band Chlorophyll Index (CIgreen), Modified Normalized Difference Vegetation Index with a blue band (mNDblue) and derived from the multi-angular images. These indices were used to estimate the N nutrition status in leaf nitrogen concentration (LNC), plant nitrogen concentration (PNC), leaf nitrogen accumulation (LNA), and plant nitrogen accumulation (PNA) of wheat canopies for a combination of treatments in N rate, variety and planting density. Their data demonstrated that the highest accuracy for single-angle images was obtained with **CIgreen for LNC** from a VZA of -60° (*R2* = 0.71, RMSE = 0.34 per cent) and PNC from a VZA of -40° (*R2* = 0.36, RMSE = 0.29 per cent). When combining an off-nadir image (-40°) and the 0° image, the accuracy of PNC estimation was substantially improved (CIred-edge: *R2* = 0.52, RMSE = 0.28 per cent). However, the use of dual-angle images did not significantly increase the estimation accuracy for LNA and PNA compared to the use of single-angle images. Their data further suggest that UAV-based multispectral camera is needed for better estimation of nitrogen concentration in wheat leaves or plants. The oblique images acquired from additional flights could be used alone or combined with the nadir-view images for improved crop N status monitoring.

In another study, Geipel at al. (2015) used fully programmable and designed as a lightweight payload for UAVs for non-destructive estimation of N in wheat. It is based on an industrial multi-sensor camera and a customizable image processing routine. The system was tested in a split fertilized N field trial at different growth stages in between the end of stem elongation and the end of anthesis. The acquired multispectral images were processed to normalized difference vegetation index (NDVI) and red-edge inflection point (REIP) orthoimages for an analysis with simple linear regression models. The best results for the estimation of above-ground biomass were achieved with the NDVI (R2 = 0.72–0.85, RMSE = 12.3 per cent –17.6 per cent), whereas N content was estimated best with the REIP (R2 = 0.58–0.89, RMSE = 7.6 per cent –11.7 per cent). Moreover, NDVI and REIP predicted grain yield at a high level of accuracy (R2 = 0.89–0.94, RMSE = 9.0 per cent –12.1 per cent). Grain protein content could be predicted best with the REIP (R2 = 0.76–0.86, RMSE = 3.6 per cent –4.7 per cent), with the limitation of prediction inaccuracies for N-deficient canopies.

Schirrmann *et al.* (2016) used low-cost UAV imagery based on an RGB consumer-level camera, were able to compute ultra-high resolution orthoimages and surface models from a cultivated 11 ha wheat field even by using only a 60 per cent image overlap. They observed the development of spatial patterns in the wheat canopy from booting to grain filling with three flight missions and were able to qualitatively assess the changes in the wheat canopy down to the individual plant level. Various biophysical parameters, *i.e.*, the leaf area index, fresh biomass, dry biomass and plant height, were highly correlated with the blue-channel-related ratios, the plant height calculated from the surface models and the plant coverage calculated from thresholding the UAV imagery. Both the image tone and the surface models derived from the UAV imagery were important to describe the spatial variability of the biophysical characteristics observed in the wheat canopies. Linear regression models for these parameters with principal components calculated from the image variables yielded R2 values between 0.70 and 0.97 for the entire data and 0.73 and 0.99 for the validation. **Schirrmann *et al.* (2016)** revealed that even under water-deficit conditions, UAV imagery can be used to estimate biophysical parameters properly. For example, deriving biomass or LAI maps from UAV imagery would help to establish irrigation systems more properly or improve plant disease forecasting.

Modeling the N content yielded only low R2 values, with the best model obtained at flowering growth stage, with an R2 of 0.65. The red image tone variables were the most important predictors because the red signal is influenced by chlorophyll absorption and chlorophyll can be related to the N content. During the later missions, the plant height and coverage derived from the UAV images became important because with further propagation of senescence, chlorophyll content decreases in the wheat leaves. When calculating the N status with the NNI, we obtained a constant error over all missions, with 0.10 to 0.11 NNI. However, the errors obtained here were comparable to other studies conducted with hyperspectral field measurements. The R2 values might be underrepresented due to the low N variability because of the soil-induced water deficit stress conditions. Therefore, the prospects for low-cost RGB UAV imagery might be limited for wheat crops under these conditions to observe the spatial variability of N and NNI because the RGB image tone was only broadly related to changes in N during the flowering growth stage. However, the combination of morphological information, calculated from surface models, and the image tone is sensible for observing the NNI because it is dependent on the biophysical and biochemical characteristics of the wheat crops. Future studies should be focused on the relationship between UAV imagery and NNI under more variable N conditions using low cost RGB cameras and hyperspectral UAV sensing should investigate wheat crops under water-deficit conditions.

5.7.4. Sunflower

Sunflower is very short duration crop but nutrient requirement should met that coincide with growth and development. Despite short duration, its nutrient requirement attains peak within 35-40 days. Therefore, accurate assessment and application of N fertilizers has to be done appropriately. Vega *et al.* (2015) have used sensors to detect the N requirement precisely. Their studies showed that the linear regressions between NDVI and grain yield, aerial biomass and nitrogen content in the biomass were significant at the 99 per cent confidence level, except during very early growth stages, whereas the time of day when the images were acquired, the classification process, and image resolution had no effect on the results. The methodology provides information that is related to crop yield from the very early stages of growth and its spatial variability within the crop field to be harvested, which can subsequently be used to prescribe the most appropriate management strategy on a site-specific basis.

Aguera *et al.* (2011) assessed the sunflower nitrogen status using multispectral images that were taken from an UAV, with optical data collected with a ground-based platform. An ADC Lite Tetracam digital cam was mounted on a md4-200 Microdrones to take pictures of a sunflower field during the crop season. ADC Lite Tetracam is a single sensor digital camera designed for capture of visible light wavelength longer than 520 nm and near-infrared wavelength up to 920 nm. The md4-200 Microdrones is an UAV which can be programmed to follow a route defined by several way-points and actions. The ground-based device was a Pacific Vision, Inc. multispectral radiometer. Four images with both systems were taken during the crop season and an index related with nitrogen crop status was calculated from them and compared in a sunflower field that had four irrigation treatments and eight nitrogen application rates, resulting in 32 plots of 7 m by 3.4 m, with a plant density of 7.1 plants m^{-2}. Calculated Normalized Difference Vegetation Index (NDVI) from both measurement systems was a good indicator of nitrogen applied, but the UAV-based system provided a better estimate than ground-based system because in the first system was possible to eliminate the soil and shadows for calculating the index.

Aguera *et al.* (2011) concluded that Images from a Tetracam on-board of an UAV can be a useful information tool for detecting nitrogen deficiencies at in-field level. Image process is easy and data can be ready in a short time. In this way, the farmer can take decisions to prescribe the most appropriate management strategy on a site-specific basis and solve nitrogen deficit during de crop season. Nevertheless, the flight autonomy is limited and the covered area is too small. Another limitation for UAV could be the weather conditions, if it is windy at a relativity low velocity, the UAV can have problems to flight. So, is necessary to develop economics UAV which have not these problems. Furthermore, the possibility to use hyperspectral sensors could to advance in the knowledge of crops nitrogen status.

5.7.5. Cotton

Cotton is a long duration crop requiring nutrient requirement coincides with the critical growth stage. The nutrient requirement increases with the progression of growth and attains the highest at the boll setting stage. In order to meet crop nutrition in accordance with the growth and requirement, scientists used spectral indices as an alternative method to determine the N requirement and rationalization of fertilizers. **Ballester *et al.* (2017)** have assessed the usefulness of a set of spectral indices obtained from an UAS for tracking spatial and temporal variability of nitrogen (N) status as well as for predicting lint yield in a commercial cotton (*Gossypium hirsutum* L.) farm. Multi-spectral images (reflectance in the blue, green, red, red edge and near infrared bands) were acquired on seven days throughout the season, from 62 to 169 days after sowing (DAS), and data were used to compute structure- and chlorophyll-sensitive vegetation indices (VIs). Above-ground plant biomass was sampled at first flower, first cracked boll and maturity and total plant N concentration (N per cent) and N uptake determined. Lint yield was determined at harvest and the relationships with the VIs explored. Results showed that differences in plant N per cent and N uptake between treatments increased as the season progressed. Early in the season, when fertilizer applications can still have an effect on lint yield, the simplified canopy chlorophyll content index (SCCCI) was the index that best explained the variation in N uptake and plant N per cent. They observed linear relationship between the VIs and both plant N per cent and N uptake at the first cracked boll and maturity, the linear regression obtained for the relationships was statistically significant, with the highest r^2 values obtained at maturity. The normalized difference red edge (NDRE) index, and SCCCI were generally the indices that best distinguished the treatments according to the N uptake and total plant N per cent. Overall, this study showed the practicality of using an UAS to monitor the spatial and temporal variability of cotton N status in commercial farms. It also illustrates the challenges of using multi-spectral information for fertilization recommendation in cotton at early stages of the crop **Ballester *et al.* (2017)** recommended monitoring SCCCI at early stages of the cotton crop, from first flower to peak bloom, when the variability can be detected and fertilizer applications can still have an effect on lint yield. Moreover, they suggested the possibility of using NDRE measurements at maturity as a potential tool to identify areas with higher or lower defoliant application needs to make prescriptive applications in order to increase harvest efficiency. SCCCI, TCARI/OSAVI and NDRE were the indices that best predicted lint yield early in the crop season (83 DAS). Results obtained from this study showed that at early stages of the crop, when cotton N demand is low and soil N availability may be enough to meet the plant N requirements, other factors may also influence plant growth and, therefore, challenge the use of structure-sensitive spectral indices such as the NDVI for predicting lint yield. The visible band index VARI performed similarly to the NDVI at predicting yield across the season, pointing out this index as an interesting option for yield prediction in cotton when only low-cost visible-band digital cameras are available.

5.7.6. Turf Grass

The turf grasses in golf course require adequate N fertilization to keep them green and aesthetic. The UAV technology appears tobe useful to determine the N content based on the spectral reflectance and fertilize accordingly using drones. Spectral reflectance data originating from UAV imagery is a valuable tool to monitor plant nutrition, reduce nitrogen (N) application to real needs, thus producing both economic and environmental benefits. Caturegli *et al.* (2016) studied the spectral reflectance of 3 different turfgrass species: *Cynodon dactylon* x *transvaalensis* (Cdxt) 'Patriot', *Zoysia matrella* (Zm) 'Zeon' and *Paspalum vaginatum* (Pv) 'Salam'. Proximity fertilized with 0 to 250 kg N ha^{-1} were created on and remote-sensed reflectance measurements were acquired using a GreenSeeker handheld crop sensor and a UAV with onboard a multispectral sensor, to determine Normalized Difference Vegetation Index (NDVI). Proximity-sensed NDVI is highly correlated with data acquired from UAV with r values ranging from 0.83 (Zm) to 0.97 (Cdxt). Relating NDVI-UAV with clippings N, the highest r is for Cdxt (0.95). The most reactive species to N fertilization is Cdxt with a clippings N per cent ranging from 1.2 per cent to 4.1 per cent. UAV imagery can adequately assess the N status of turfgrasses and its spatial variability within a species, so for large areas, such as golf courses, sod farms or race courses, UAV acquired data can optimize turf management.

Similar studies were taken up Bermuda grass and tall fescue to develop an economic hand device for N estimation. In this study, NDVI acquired by a ground-based handheld crop sensor and by a multispectral camera mounted on board a UAV has been compared with DGCI calculated from images taken with a commercial digital camera on board a UAV, trying to quantify the colour of turfgrass that had received different nitrogen (N) rates. The study came up with an affordable easy-to-use tool evaluating the relationship among NDVI, DGCI and leaf nitrogen content on turfgrass.

5.7.7. Trees

The variable fertilizer applicator is also tested in tree species in Canada. Felderhof and Gillieson (2011) reported that the aerial images obtained using an UAV were used to create a classified image showing tree canopy health in a **macadamia orchard**. The resulting map was used to modify management of a macadamia plantation. Vegetation indices, principally the Canopy Chlorophyll Content Index (CCCI), derived from both UAS and WorldView2 satellite imagery, were compared and correlated with spectral radiometry and leaf nitrogen levels determined by field sampling. Classified CCCI images from both sensor types were integrated into farm management software (PAM Ultracrop) and processed into a suitable format for driving a GPS-controlled fertilizer spreader for more effective control of nitrogen application rates. Applying fertilizer at a variable rate according to tree health will result in cost savings to the industry and potentially increase production.

The extensive studies on drones based fertilization have clearly demonstrated that drones with sensors can be effectively used for rationalizing fertilizer use for crops especially nitrogen. Major part of the studies across the globe worked primarily on nitrogen as it is universally deficient and the gradient of N deficiency closely coincided with the greenness of leaves. The spectral signature is quite distinct and easily interpreted. Despite the fact that the technology is proved effective for N management, ore research is required for the assessment of P, K and other secondary and micronutrients.

Chapter 6

Drones for Irrigation Management

Irrigation is one of the most critical and demanding inputs in agriculture deciding the agricultural productivity of any crops. The impact will be so severe in rainfed systems where the crops are cultivated with rainfall and uncertainty and erratic rainfall cause unpredictability of sustained production and rural livelihood. The per capita availability of irrigation water in India has reduced drastically over the years. Water availability per person is dependent on population of the country and for India, per capita water availability in the country is reducing due to increase in population. The average annual per capita water availability in the years 2001 and 2011 was assessed as **1816** cubic meters and **1545** cubic meters respectively which may further reduce to **1486** cubic meters in the year 2021. Within the country, Tamil Nadu is the most water starving States next to Rajasthan. In Tamil Nadu State, the per capita water availability is 900 cubic metre and further declining over the years. The country has anchored a concept of "**more crop per drop of water**" to manage the available water best way possible. In this context, use of remote sensing and sensor technologies may be beneficial to detect moisture in soil and plant more precisely so as to rationalize the amount of water available for crop production.

The use of UAVs in irrigation management is primarily focused on orchards and perennials and rarely studied in field crops like oats, cotton and groundnut. In all the cases, the leaf moisture status is correlated with imageries and adjudge irrigation water requirement.

6.1. Barley

Barley is one of the model system extensively studied for UAV enabled irrigation management. Hoffmann *et al.* (2016) investigated whether a water deficit index

(WDI) based on imagery from unmanned aerial vehicles (UAVs) can provide accurate crop water stress maps at different growth stages of barley and in differing weather situations. Data from both the early and late growing season are included to investigate whether the WDI has the unique potential to be applicable both when the land surface is partly composed of bare soil and when crops on the land surface are senescing. The WDI differs from the more commonly applied crop water stress index (CWSI) in that it uses both a spectral vegetation index (VI), to determine the degree of surface greenness, and the composite land surface temperature (LST) (not solely canopy temperature). Lightweight thermal and RGB (red–green–blue) cameras were mounted on a UAV on three occasions during the growing season 2014, and provided composite LST and color images, respectively. From the LST, maps of surface-air temperature differences were computed. From the color images, the normalized green–red difference index (NGRDI), constituting the indicator of surface greenness, was computed. Advantages of the WDI as an irrigation map, as compared with simpler maps of the surface-air temperature difference, are discussed, and the suitability of the NGRDI is assessed. Final WDI maps had a spatial resolution of 0.25 m. It was found that the UAV-based WDI is in agreement with measured stress values from an eddy covariance system. Further, the WDI is especially valuable in the late growing season because at this stage the remote sensing data represent crop water availability to a greater extent than they do in the early growing season, and because the WDI accounts for areas of ripe crops that no longer have the same need for irrigation. WDI maps can potentially serve as water stress maps, showing the farmer where irrigation is needed to ensure healthy growing plants, during entire growing season.

Water Deficit Index (WDI) is calculated as defined by Moran *et al.* (1994) as:

$$WDI = 1 - \lambda Eact \;/\; \lambda Epot = (TS-TA)min-(TS-TA)mes \;/\; (TS-TA)min-(TS-TA)max$$

λEact is the actual latent heat flux density (W m^{-2})

λEpot is the potential latent heat flux density (W m^{-2})

TS is the composite LST land surface temperature (^{o}C),

TA is the air temperature (^{o}C)

Subscripts 'min', 'max' and 'mes' refer to minimum, maximum and measured.

Hoffmann *et al.* (2016) have compared the visual comparison of NGRDI maps and corresponding orthophotos (Figures 6.1-6.3, panels A and C) and confirmed that the NGRDI can be used to quantify greenness from the RGB data. Yellowing of barley plants (Figures 6.2-6.3) are also exhibited. The green stripes along the edges of the fields in the orthophoto and a northeast–southwest-orientated stripe of green vegetation in the field north of the gravel road in Figure 6.1c are translated into brighter areas in Figure 6.1a (purple lines), indicating more green vegetation and a higher canopy cover. Comparing Figure 6.2a with c and Figure 6.3a with c shows that the NGRDI is also able to quantify the degree of crop ripeness. Areas with ripe crops attain lower (darker) NGRDI values (areas highlighted with blue). The same NGRDI response is observed in areas with bare soil, and therefore the NGRDI responses to ripe crops and bare soil are indeed similar.

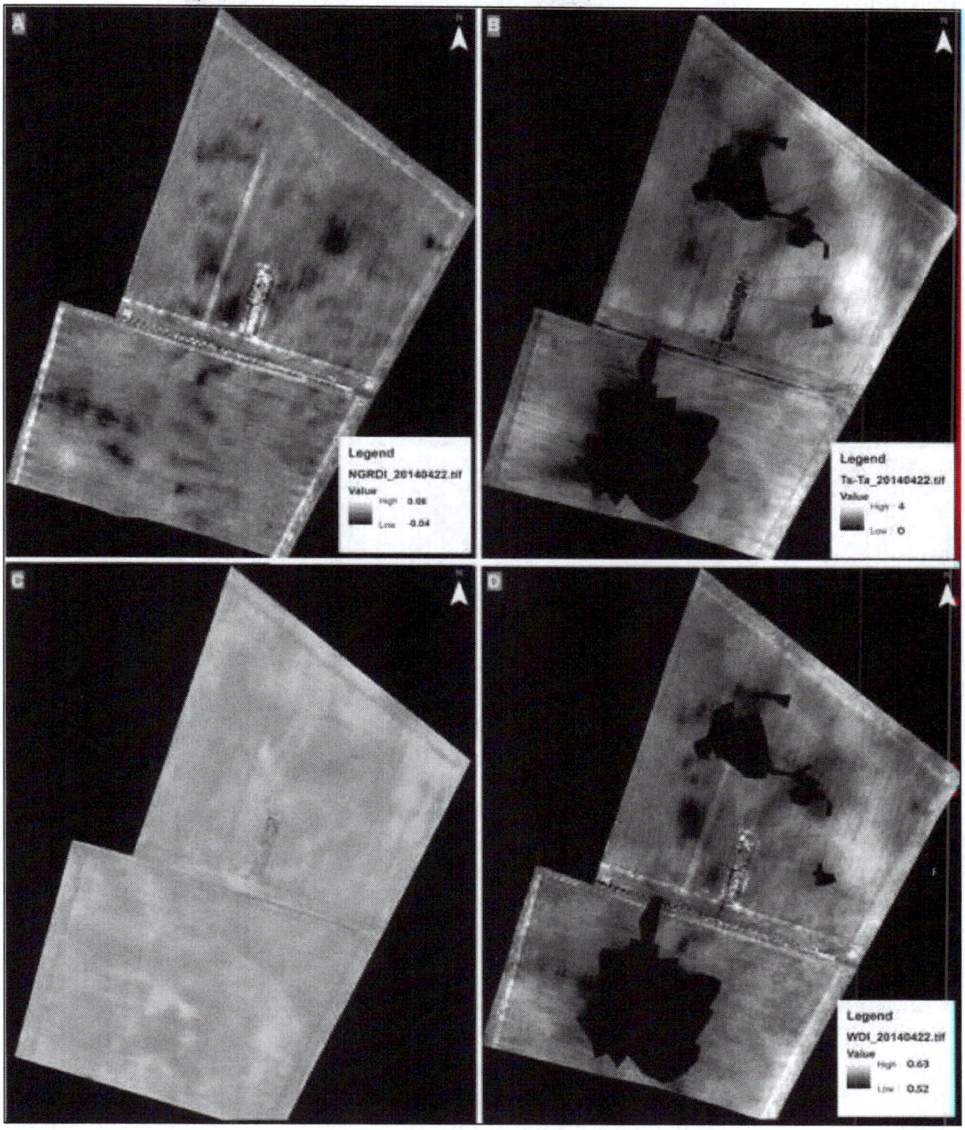

Figure 6.1: Imagery and Indices from April 22. (a) NGRDI; (b) TS-TA; (c) Orthophoto (RGB data) and D) WDI. Black areas (no data) within the field on maps (b) and (d) are due to insufficient texture in images collected in these areas, which meant that the images could not be aligned with the Structure-from-Motion algorithm. (*Source*: Hoffmann *et al.,* 2016).

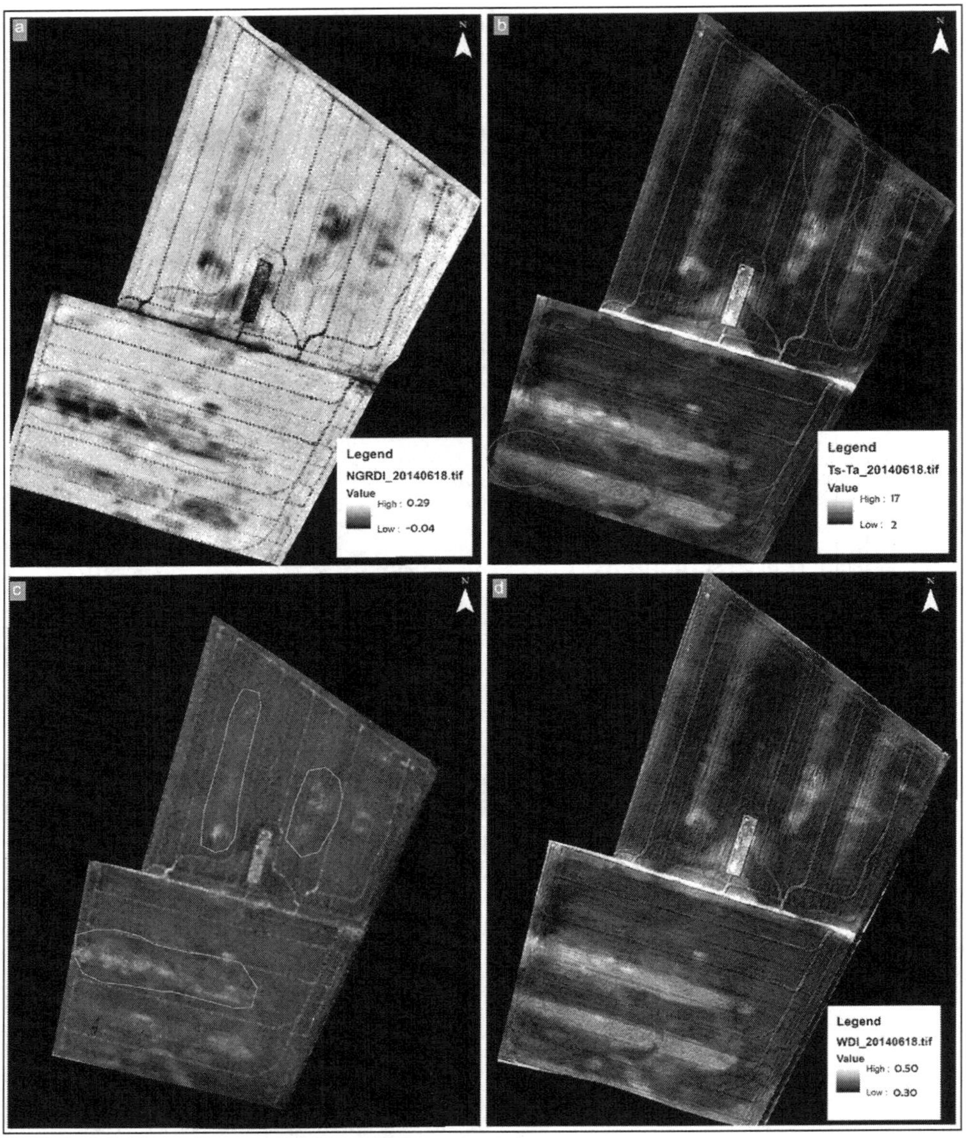

Figure 6.2: Imagery and Indices from 18 June. (a) NGRDI; (b) TS - TA; (c) Orthophoto (RGB data) and (d) WDI. Red circles in (b) are explained in Sect. 4.2. Areas highlighted with blue are examples of locations with ripe crops (*Source*: Hoffmann *et al.,* 2016).

The NGRDI map from **18 June** was compared with approximately 1300 spatially distributed measurements of NDVI from 22 June, and a correlation coefficient (r) of 0.56 was achieved ($p < 0.001$). The offset of 4 days between NDVI and NGRDI measurements was expected to introduce small differences in the detected greenness in areas where vegetation is near senescence. The correlation might have been stronger had the dates corresponded. However, it suggests a medium–strong association.

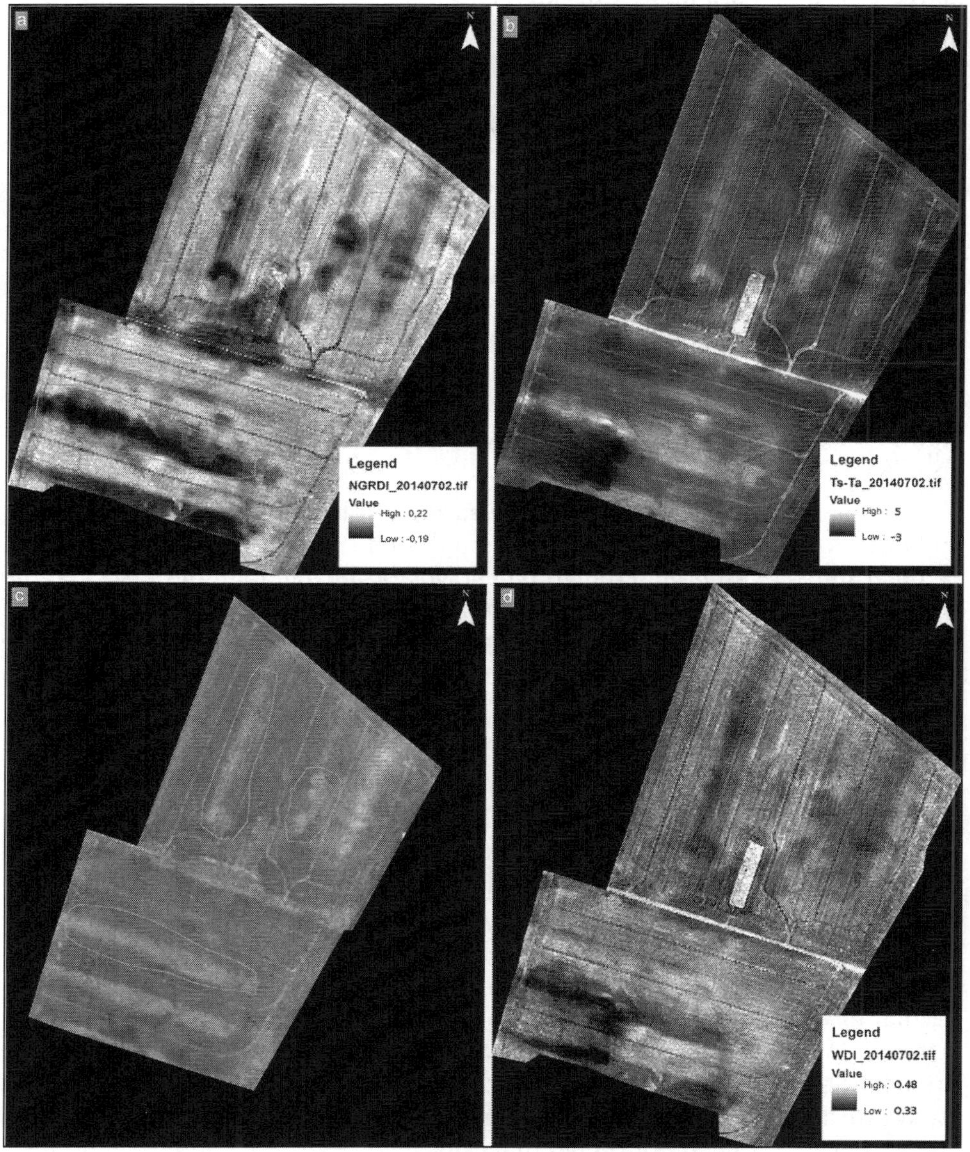

Figure 6.3: Imagery and Indices from 2 July. (a) NGRDI; (b) TS - TA; (c) Orthophoto (RGB data) and (d) WDI. Areas highlighted with blue are examples of locations with ripe crops (*Source*: Hoffmann *et al.,* 2016).

Comparison of Figure 6.2b, c and 7c reveals that thermal UAV data contain more information than it is possible to see simply by looking at the RGB images, *i.e.*, more than what can be seen with the naked eye. White areas highlighted by red circles (Figure 6.2b) indicate higher temperatures and higher soil water deficits that cannot be detected in the orthophoto (Figure 6.2c). Crops in the areas highlighted with red circles in the 18 June map become yellow by **2 July** (Figure 6.3c). The temperature data from 18 June thus predict where crops on 2 July will become prematurely ripe as a consequence of being water stressed and receiving insufficient irrigation.

Figures 6.1–3 (panels a, b and d) show that variations in both TS - TA and NGRDI maps are reflected in the WDI maps. This can most clearly be seen on 22 April (Figure 6.1a, b and d) where variations in NGRDI and TS - TA maps deviate most from each other. On 18 June, areas with high TS - TA and green vegetation (high NGRDI values) are translated into areas with higher water deficits (Figure 6.3d). The same areas have high TS - TA on 2 July. However, the increasing incidence of ripe/prematurely ripe crops in the same areas now translates into lower WDI values. When areas are detected as those with ripe/prematurely ripe crops, the TS - TA has to

In this study, the UAV-based WDI was applied to barley fields in April, June and July (2014), to investigate whether crop water deficits at subfield scale could be determined at different crop growth stages. Data from both the early and the late growing season were investigated to assess whether the WDI has the unique potential to be applicable both when the land surface is partly composed of bare soil and when crops on the land surface are senescing.

Hoffman *et al.* (2016) found that the WDI maps determined accurate absolute water stress values and variations within the barley fields, in agreement with measured stress values from the eddy covariance tower, at different growth stages. The WDI index provides very accurate estimates of crop water status, and therefore the WDI maps provide precise irrigation maps. WDI maps are most valuable in the late growing season. This implies that the WDI accounts for areas with ripe and prematurely ripe crops that no longer need large volumes of irrigation. The robustness of the WDI during different growth stages emphasizes its added utility compared with the more commonly used CWSI. Further, the WDI has the potential to become an efficient and powerful irrigation tool in areas where overcast weather is common. We also found that the surface-air temperature difference maps alone can predict where crops will become prematurely ripe with insufficient irrigation. The study also demonstrated that a lightweight UAV system, a consumer-grade camera and an uncalibrated and uncooled thermal camera can be combined to produce accurate maps of crop water stress. In this way, reliance on camera calibration and costly multispectral cameras can be reduced. The WDI response in the late growing season was crop-color dependent, and studies applying the setup we have presented to other crop types are needed in order to confirm the general added utility of the WDI. The WDI can also be used in other cereals as well but there is a need for validation.

6.2. Grapes

Irrigation scheduling is critical for vineyards and olive orchards, since it directly affects yield and fruit composition. Poblete-Echeverrýìa *et al.* (2016) studied the Regulated deficit irrigation (RDI) strategies. Canopy temperature (Tc) has been shown to be an accurate indicator of plant water stress. The practical application of RDI requires an accurate knowledge of plant water status, which is commonly achieved using a pressure chamber (Scholander type). This method has demonstrated to be an accurate indicator of water status in olive trees (*e.g.*, Fernaìndez *et al.*, 2006; Martýìn- Vertedor *et al.*, 2011) and grapevines (Girona *et al.*, 2006). However, the use of pressure chamber for measuring water potential is impractical for precision agriculture since a high number of measurements are required to have a good spatial representation (Jones *et al.*, 2004). Stomata conductance (gs) is another technique used to quantify water status. This parameter is sensitive to water stress (Medrano *et al.*, 2002), however, has several disadvantages that limit its practical use for irrigation scheduling (Fuentes *et al*, 2012). Nowadays, canopy temperature (Tc) has gained attention as a non-invasive indicator of plant water stress since plants undergoing water stress increase their temperature (Jones *et al.*, 2002; Wang *et al.*, 2010). Useful information on representative canopy temperature can be derived from infrared thermal images and the temperatures are highly correlated with stomatal conductance (Fuentes *et al.*, 2012) and stem water potential (Wang *et al.*, 2010). To use infrared thermal images, however, it is critical to identify only leaf temperature from the full image to avoid over- or under-estimations of Tc due to non-leaf material inclusion in the analysis (Fuentes *et al.*, 2012). Nowadays, two methods are used to acquire infrared thermal images in row crops: i) lateral images captured by handheld infrared cameras and ii) nadir-view images obtained using a telescopic pole or unmanned aerial vehicles (UAVs). The two methods have showed different results when compared with field water status measures. Therefore, the objective of this study was to evaluate the accuracy of water stress detection based on aerial and terrestrial infrared thermography for a vineyard and an olive orchard.

Poblete-Echeverrýìa *et al.* (2016) evaluated the accuracy of water stress detection based on aerial and terrestrial infrared thermography for a vineyard and an olive orchard. Lateral infrared thermal images were obtained using a handheld infrared camera and nadir-view infrared thermal images were obtained using an unmanned aerial vehicle (UAV). In addition, measurements of midday stem water potential (Ψstem) from olives trees (*Olea europaea* L. 'Arbequina') and grapevines (*Vitis vinifera* L. 'Carménère') were taken under different RDI strategies during the 2013-2014 growing season. The image analysis was performed using a customized code written in Mathlab® defining thresholds to exclude non-plant elements. Results showed that the use of infrared thermal images aided in recognizing the differences in the water availability for irrigation treatments when the plants were under stress conditions. This study was developed to evaluate the accuracy of water stress detection based on baerial and terrestrial infrared thermography for a vineyard and an olive orchard. The canopy minus air temperature difference Tc-Ta obtained from aerial and terrestrial methods, were related to differences in water potential for different irrigation treatments when the plants were under stress conditions. Also, results showed that by the use of infrared thermal analysis it is possible to detect water stress in olive trees and grapevines by identifying critical thresholds between non-stress and stress situations as a function of the differential temperature.

Espinoza *et al.* (2017) analyzed the multispectral and thermal infrared images acquired from an unmanned aerial vehicle (UAV) to evaluate the applicability of the data in the assessment of variants of subsurface irrigation configurations. The study was carried out in a Cabernet Sauvignon orchard located near Benton City, Washington. Plants were subsurface irrigated at a 30, 60, and 90 cm depth, with 15 per cent, 30 per cent, and 60 per cent irrigation of the standard irrigation level as determined by the grower in commercial production management. Half of the plots were irrigated using pulse irrigation and the other half using continuous irrigation techniques. The treatments were compared to the control plots that received standard surface irrigation at a continuous rate. The results showed differences in fruit yield when the control was compared to deficit irrigated treatments (15 per cent, 30 per cent, 60 per cent of standard irrigation), while no differences were found for comparisons of the techniques (pulse, continuous) or depths of irrigation (30, 60, 90 cm). Leaf stomatal conductance of control and 60 per cent irrigation treatments were statistically different compared to treatments receiving 30 per cent and 15 per cent irrigation. The normalized difference vegetation index (NDVI), green normalized difference vegetation index (GNDVI), and canopy temperature were correlated to fruit yield and leaf stomatal conductance. Significant correlations ($p < 0.01$) were observed between NDVI, GNDVI, and canopy temperature with fruit yield (Pearson's correlation coefficient, $r = 0.68$, 0.73, an 0.83, respectively), and with leaf stomatal conductance ($r = 0.56$, 0.65, and 0.63, respectively) at 44 days before harvest. This study demonstrates the potential of using low-altitude multispectral and thermal imagery data in the assessment of irrigation techniques and relative degree of plant water stress. In addition, results provide a feasibility analysis of our hypothesis that thermal infrared images can be used as a rapid tool to estimate leaf stomatal conductance, indicative of the spatial variation in the vineyard. This is critically important, as such data will provide a near real-time crop stress assessment for better irrigation management/scheduling in wine grape production.

Santesteban *et al.* (2017) used thermal imaging as a readily usable tool for crop agricultural water management, since it allows a quick determination of canopy surface temperature that, as linked to transpiration, can give an idea of crop water status. In the last years, the resolution of thermal imaging systems has increased and its weight decreased, fostering their implementation on Unmanned Aerial Vehicles (UAV) for civil and agricultural engineering purposes. This approach would overcome most of the limitations of on sitethermal imaging, allowing mapping plant water status at either field or farm scale, taking thus into account the naturally existing or artificially induced variability at those scales. The aim of this work was to evaluate to which extent high-resolution thermal imaging allows evaluating the instantaneous and seasonal variability of water status within a vineyard. The novelty and significance of our approach is that the specifically designed and build unmanned aerial vehicle (UAV) provided very high-resolution imaging (pixel <9 cm), and that it was used at a commercially relevant acreage (7.5 ha). This set-up was used to obtain Crop Water Stress Index (CWSI) from thermal images in a clear-sky day. CWSI values were and compared to stem water potential (SWP) and stomatal conductance (gs) measured at 14 sampling sites across the vineyard at the moment when images where acquired. In order to evaluate the potential of CWSI acquired in a single day to estimate within-vineyard patterns of variation in water status, a spatial modeling approach was used. CWSI correlated well with SWP and gs at the moment of image acquisition, showing to have a great potential to monitor instantaneous variations in water status within a vineyard. The information provided by thermal images proved to be relevant at a seasonal scale as well, although it did not match seasonal trends in water status but mimicked other physiological processes occurring during ripening. Therefore, if a picture of variations in water status is required, it would be necessary to acquire thermal images at several dates along the summer.

Poblete *et al.* (2017) evaluated the midday stem water potential (SWP). However, this measurement is acquired on a per plant basis and does not account for the assessment of vine water status spatial variability. The use of multispectral cameras mounted on unmanned aerial vehicle (UAV) is capable to capture the variability of vine water stress in a whole field scenario. It has been reported that conventional multispectral indices (CMI) that use information between 500–800 nm, do not accurately predict plant water status since they are not sensitive to water content. The objective of this study was to develop artificial neural network (ANN) models derived from multispectral images to predict the Ystem spatial variability of a drip-irrigated Carménère vineyard in Talca, Maule Region, Chile. The coefficient of determination (R2) obtained between ANN outputs and ground-truth measurements of Ystem were between 0.56–0.87, with the best performance observed for the model that included the bands 550, 570, 670, 700 and 800 nm. Validation analysis indicated that the ANN model could estimate Ystem with a mean absolute error (MAE) of 0.1 MPa, root mean square error (RMSE) of 0.12 MPa, and relative error (RE) of 9.1 per cent. For the validation of the CMI, the MAE, RMSE and RE values were between 0.26–0.27 MPa, 0.32–0.34 MPa and -24.2–25.6 per cent, respectively.

Poblete *et al.* (2017) evaluated the midday stem water potential (SWP) and thermal infrared (TIR) imaging for crop water stress index (CWSI) have been used to assess plant water stress on a vine-by-vine basis without considering the spatial variability. Unmanned Aerial Vehicle (UAV)-borne TIR images are used to assess the canopy temperature variability within vineyards that can be related to the vine water status. Nevertheless, when aerial TIR images are captured over canopy, internal shadow canopy pixels cannot be detected, leading to mixed information that negatively impacts the relationship between CWSI and SWP. This study proposes a methodology for automatic co-registration of thermal and multispectral images (ranging between 490 and 900 nm) obtained from a UAV to remove shadow canopy pixels using a modified scale invariant feature transformation (SIFT) computer vision algorithm and Kmeans++ clustering. Our results indicate that our proposed methodology improves the relationship between CWSI and SWP when shadow canopy pixels are removed from a drip-irrigated Cabernet Sauvignon vineyard. In particular, the coefficient of determination (R2) increased from 0.64 to 0.77. In addition, values of the root mean square error (RMSE) and standard error (SE) decreased from 0.2 to 0.1 MPa and 0.24 to 0.16 MPa, respectively. Finally, this study shows that the negative effect of shadow canopy pixels was higher in those vines with water stress compared with well-watered vines.

Romero *et al.* (2018) provided a fast and reliable alternative for traditional in situ water status measurement in vineyards. Several vegetation indices (VIs) derived from aerial multispectral imagery were tested to estimate midday stem water potential (Ψstem) of grapevines. The experimental trial was carried out in a vineyard in the Shangri-La region, located in Yunnan province in China. Statistical methods and machine learning algorithms were used to evaluate the correlations between Ψstem and VIs. Results by simple regression between VIs individually and Ψstem showed no significant relationships, with coefficient of determination (R2) for linear fitting smaller than 0.3 for almost all the indices studied, except for the Optimal Soil Adjusted Vegetation Index (OSAVI); R2=0.42 with statistical significance ($p \leq 0.001$). However, results from a model obtained by fitting using Artificial Neural Network (ANN), using all VIs calculated as inputs and real Ψstem from plants within the study site (n=90) as targets (Model 1), showed high correlation between the estimated water potential through ANN (Ψstem ANN) and the actual measured Ψstem. Training, validation and testing data sets presented individual correlations of R=0.8, 0.72 and 0.62 respectively. The models obtained from the study site were then applied to a wider area from the vineyard studied and compared to further Ψstem measured obtained from different sites (n=23) showing high correlation values between Ψstem ANN and real Ψstem (R2=0.83; slope=1; $p \leq 0.001$). Finally, a pattern recognition ANN model (Model 2) was developed for irrigation scheduling purposes using the same Ψstem measured in the study site as inputs and with the following thresholds as outputs: Ψstem below –1.2 MPa considered as severe water stress (SS), Ψstem between –0.8 to –1.2 MPa as moderate stress (MS) and Ψstem over–0.8 MPa with no water stress (NS). This model can be applied to analyze on a plant by plant basis to identify sectors of stress within the vineyard for optimal irrigation management and to identify spatial variability within the vineyards.

6.3. Orchards

Gonzalez-Dugo *et al.* (2013) assessed the heterogeneity in water status in a commercial orchard, as a prerequisite for precision irrigation management. Remote sensing-derived indicators could be suitable for mapping water stress over large areas, and recent studies have demonstrated that high resolution airborne thermal imagery enables the assessment of discontinuous canopies as pure tree crowns can be targeted, thus eliminating the background effects. Airborne campaigns were conducted over a drip-irrigated commercial orchard in Southwestern Spain composed of five different orchard tree crops. An unmanned aerial vehicle with a thermal camera onboard was flown three times during the day on 8 July 2010, at 9, 11 and 13 h (local time). Stem water potential was measured at the same time of the flights. In some irrigation units, irrigation was stopped prior to the measurement date to induce water deficits for comparative purposes. Several approaches for using the thermal data were proposed. Daily evolution of the differential between canopy and air temperature (Tc - Ta) was compared to tree water status. The slope of the evolution of Tc - Ta with time was well correlated with water status and is proposed as a novel indicator linked with the stomatal behavior. The Crop Water Stress Index (CWSI) was calculated with the temperature data from the 13.00 h flight using an empirical approach for defining the upper and lower limits of Tc - Ta.

Jackson and coworkers proposed the concept of **Crop Water Stress Index (CWSI),** defined as the difference between air (Ta) and canopy temperature (Tc), normalized for the evaporative demand as determined by means of a lower limit (the case of a canopy transpiring at its potential rate) and an upper limit (a non-transpiring canopy), as shown in Eq. (1):

$$\textbf{CWSI} = (T_c - T_a) - (T_c - T_a)\,LL / (T_c - T_a) - (T_c - T_a)\,UL$$

Where, **UL** and **LL** are the upper and lower limits, respectively.

The assessment of variability in water status was also performed using differences in relative canopy temperatures. Ample variability was detected among and within irrigation units, demonstrating that the approach proposed was viable for precision irrigation management. The assessment led to the identification of water-stressed areas, and to the definition of threshold CWSI values and associated risks. Such thresholds may be used by growers for irrigation management based on crop developmental stages and economic considerations.

Indicators derived from thermal imagery acquired from an UAV over a commercial orchard have been shown to accurately describe the spatial variability in crop water status. While physiologically-based indicators can be measured in a limited number of trees, thermal imagery provides information for whole fields and if the resolution is sufficient, as shown in this study, it allows the mapping of an orchard on a tree by tree basis. It becomes thus a valuable tool for water management in precision agriculture and deficit irrigation strategies. The daily course of thermal-derived indicators was also analyzed. The slope of the evolution of Tc - Ta with time was well correlated with water status and is proposed as a

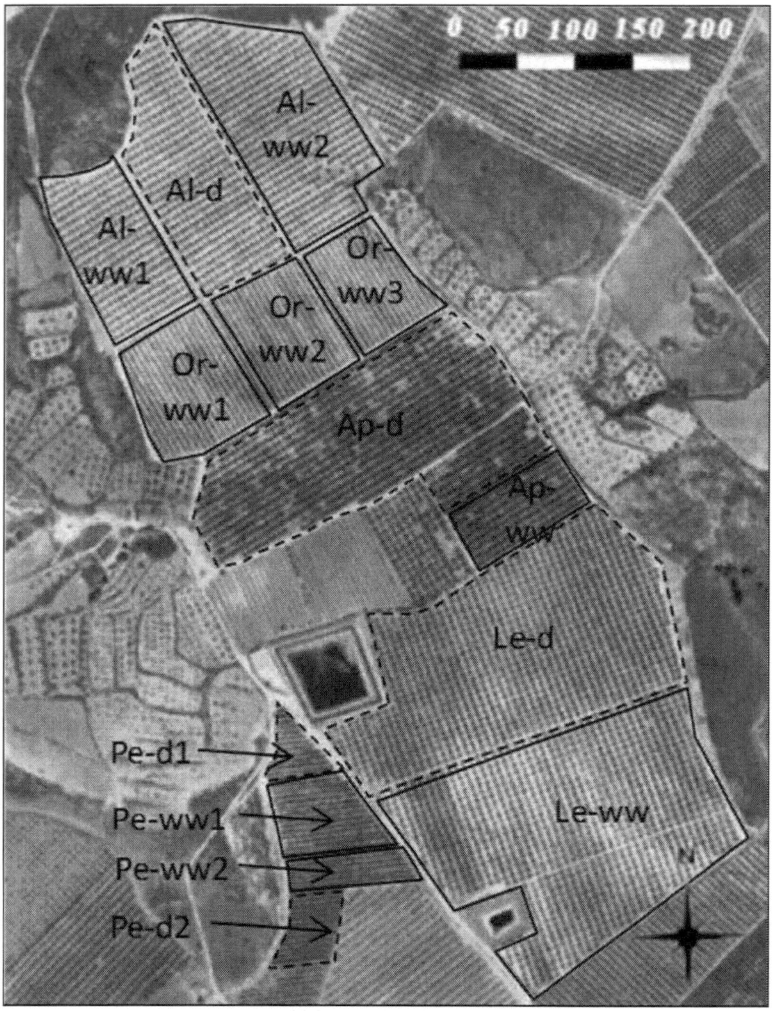

Fig. 6.4: Aerial view of the orchard and its irrigation management units. Al, Or, Ap, Le and Pe correspond to almond, orange, apricot, lemon and peach, respectively. Solid line indicate the plots that were fully irrigated (-ww plots), while dotted lines correspond with the plots where irrigation was withheld prior to the flights(-d plots).

promising indicator to track water status. Its main advantage relative to point measurements of temperature is that it encompasses the stomatal behavior for longer periods of time. Moreover, the indicator behaved similarly for the three Prunus species. More studies under different conditions are required to assess the robustness of this approach.

The analysis of the variability of Tc demonstrated that data obtained at 13.00 were the most suitable for describing the spatial variability of canopy temperature;

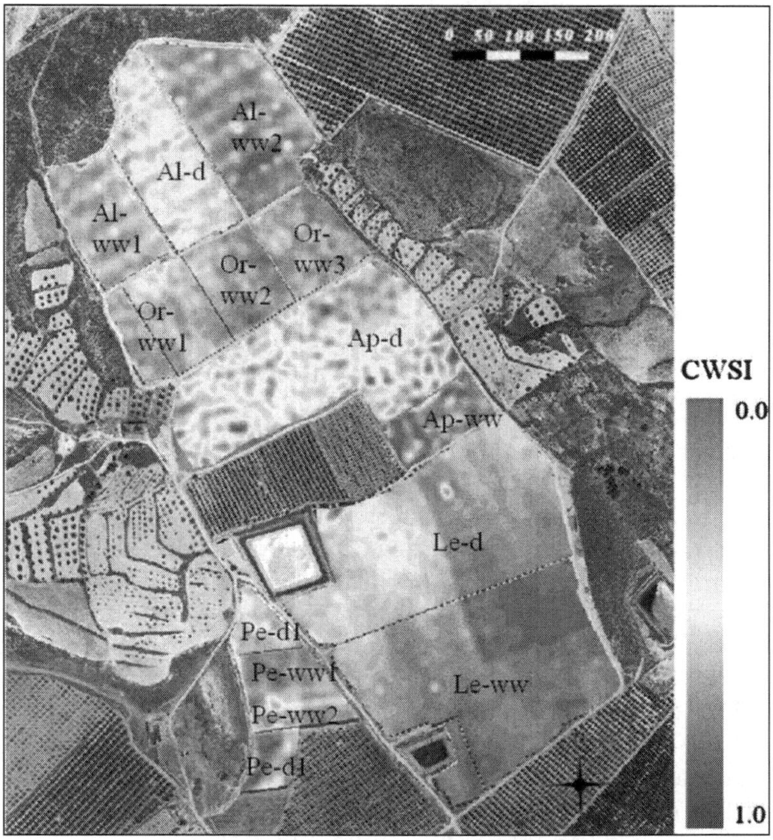

Fig. 6.5: Map of the CWSI of the Orchard, Showing the Scale from Minimum to Maximum CWSI (*Source*: Gonzalez-Dugo *et al.,* 2013).

thus, CWSI was calculated for this time. The methodology proposed here enabled the determination of CWSI for the whole orchard, tree by tree, based on images of individual flights performed over an orchard where five tree species were grown commercially. Future work should determine CWSI thresholds for different species and phenological stages leading to a quantitative framework suitable for assisting in irrigation management decisions. The CWSI thresholds would be set based not only on physiological and horticultural responses to water stress but on economic considerations as well.

In another study, **Park *et al.* (2017)** monitored water status of nectarine and peach orchards using high-resolution thermal imagery acquired by an unmanned aerial vehicle (UAV) The study site was classified into sub-regions based on crop properties, such as cultivars and tree training systems. In order to enhance the accuracy of the mapping, edge extraction and filtering were conducted prior to the probability modelling employed to obtain crop-property-specific ('adaptive' hereafter) lower and higher temperature references (Twet and Tdry respectively). Direct measurements of stem water potential (SWP) and stomatal conductance (gs) were collected concurrently with UAV remote sensing and used to validate the thermal index as crop biophysical parameters. The adaptive crop water stress index (CWSI) presented a better agreement with both ystem and gs with determination coefficients (R^2) of 0.72 and 0.82, respectively, while the conventional CWSI applied by a single set of hot and cold references resulted in biased estimates with R^2 of 0.27 and 0.34, respectively. Using a small number of ground-based measurements of SWP, CWSI was converted to a high-resolution SWP map to visualize spatial distribution of the water status at field scale. The results have important implications for the optimal management of irrigation for crops.

6.4. Citrus

Stagakis *et al.* (2012) monitored the water status and the assessment of the effect of stress on citrus fruit quality using structural and physiological remote sensing indices. Four flights were conducted over a citrus orchard in 2009 using an unmanned aerial vehicle (UAV) carrying a multispectral camera with six narrow spectral bands in the visible and near infrared. Physiological indices such as the Photochemical Reflectance Index (PRI570), a new structurally robust PRI formulation that uses the 515 nm as the reference band (PRI515), and a chlorophyll ratio (R700/R670) were compared against the Normalized Difference Vegetation Index (NDVI), Renormalized Difference Vegetation Index (RDVI) and Modified Triangular Vegetation Index (MTVI) canopy structural indices for their performance in tracking water status and the effects of sustained water stress on fruit quality at harvest. The irrigation setup in the commercial orchard was compared against a treatment scheduled to satisfy full requirements (based on estimated crop evapotranspiration) using two regulated deficit irrigation (RDI) strategies. The water status of the trees throughout the experiment was monitored with frequent field measurements of stem water potential (Wx), while titratable acidity (TA) and total soluble solids (TSS) were measured at harvest on selected trees from each irrigation treatment. The high spatial resolution of the multispectral imagery (30 cm pixel size) enabled identification of pure tree crown components, extracting the tree reflectance from shaded, sunlit and aggregated pixels. The physiological and structural indices were then calculated from each tree at the following levels: (i) pure sunlit tree crown, (ii) entire crown, aggregating the within-crown shadows, and (iii) simulating a lower resolution pixel, including tree crown, sunlit and shaded soil pixels. The resulting analysis demonstrated that both PRI formulations were able to track water status, except when water stress altered canopy structure. In such cases, PRI570 was more affected than PRI515 by the structural changes caused by sustained water stress throughout the season. Both PRI formulations were proven to serve as pre-visual water stress indicators linked to fruit quality TSS and TA parameters ($r2 = 0.69$ for PRI515 vs TSS; $r2 = 0.58$ vs TA). In contrast, the chlorophyll (R700/R670) and structural indices (NDVI, RDVI, MTVI) showed poor relationships with fruit quality and water status levels ($r2 = 0.04$ for NDVI vs TSS; r2 = 0.19 vs TA). The two PRI formulations showed strong relationships with the field-measured fruit quality parameters in September, the beginning of stage III, which appeared to be the period most sensitive to water stress and the most critical for assessing fruit quality in citrus. Both PRI515 and PRI570 showed similar performance for the two scales assessed (sunlit crown and entire crown), demonstrating that within crown component separation is not needed in citrus tree crowns where the shaded vegetation component is small. However, the simulation conducted through spatial resampling on tree + soil aggregated pixels revealed that the physiological indices were highly affected by soil reflectance and between-tree shadows, showing that for TSS vs PRI515 the relationship dropped from r2 = 0.69 to r2 = 0.38 when aggregating soil + crown components. This work confirms a previous study that demonstrated the link between PRI570, water stress, and fruit quality, while also making progress in assessing the new PRI formulation (PRI515), the within-crown shadow effects on the physiological indices, and the need for high resolution imagery to target individual tree crowns for the purpose of evaluating the effects of water stress on fruit quality in citrus.

Zarco-Tejada *et al.* (2012) evolved a new strategy to measure the water stress in citrus orchards. They measured chlorophyll fluorescence and Photochemical Reflectance Index (PRI) data, seasonal time-series of crown temperature and PRI, and high-resolution airborne imagery. The work was conducted in an orchard where a regulated deficit irrigation (RDI) experiment generated a gradient in water stress levels. Stomatal conductance (Gs) and water potential (Ψ)were measured over the season on each treatment block. The airborne data consisted on thermal and hyperspectral imagery acquired at the time of maximum stress differences among treatments, prior to the re-watering phase, using a miniaturized thermal camera and a micro-hyperspectral imager on board an unmanned aerial vehicle (UAV). The hyperspectral imagery was acquired at 40 cm resolution and 260 spectral bands in the 400-885 nm spectral range at 6.4 nm full width at half maximum (FWHM) spectral resolution and 1.85 nmsampling interval, enabling the identification of pure crowns for extracting radiance and reflectance hyperspectral spectra from each tree. The FluorMOD model was used to investigate the retrieval of chlorophyll fluorescence by applying the Fraunhofer Line Depth (FLD) principle using three spectral bands (FLD3), which demonstrated that fluorescence retrieval was feasible with the configuration of the UAV micro-hyperspectral instrument flown over the orchard. Results demonstrated the link between seasonal PRI and crown temperature acquired from instrumented trees and field measurements of stomatal conductance and water potential. The sensitivity of PRI and Tc–Ta time-series to water stress levels demonstrated a time delay of PRI vs Tc–Ta during the recovery phase after re-watering started. At the time of the maximum stress difference among treatment blocks, the airborne imagery acquired from the UAV platform demonstrated that the crown temperature yielded the best coefficient of determination for Gs (r^2=0.78; pb0.05) and Ψ (r^2=0.34; pb0.001). Among the narrow-band indices calculated, the PRI515 index (reference band=515 nm) obtained better results than PRI570, with r2=0.59 (pb0.01) for Gs, and r^2=0.38

Table 2: Monitoring of Irrigation Water Status in Crops using Drones

Sl.No.	*Crops*	*Sensors*	*UAV Types*	*Estimated Variables*	*Reference Variables*	*Model Type*	*Reference*
1	Groundnut	Multispectral	Rotary	NDVI	Visual wilting score	Linear regression	Balota *et al.*, 2017
2	Cotton	Thermal	Fixed wing	TIR emittance	Soil water content	Linear regression	Sullivan *et al.*, 2007
3	Peach orchard	Multispectral, thermal	Rotary	PRI	Stomatal conductance	Linear regression	Berni *et al.*, 2009
4	Olive orchard	Hyperspectral, thermal	Fixed wing	Tc -Ta, CSWI, CF	Stomatal conductance	Linear regression	Calderón *et al.*, 2013
5	Apple orchard	Multispectral, thermal, RGB	Rotary	Tc - Ta, WDI	Soil water potential	Direct comparison	Gómez *et al.*, 2017
6	Citrus orchard	Hyperspectral	Fixed wing	PRI	Water content	Linear regression	Delalieux *et al.*, 2014
7	Vineyard	Multispectral, thermal	Rotary	NDVI, GNDVI, Tc	Stomatal conductance	Linear regression	Espinoza *et al.*, 2017
8	Grapes	Thermal	Fixed wing	CWSI	Leaf water potential	Linear regression	Bellvert *et al.*, 2014
9	Almond orchard	Multispectral, RGB	Rotary	NDVI	Stem water potential	Linear regression	Zhao *et al.*, 2015 Zhao *et al.*, 2017
10	Orange orchard	Multispectral	Rotary	PRI	Stem water potential	Linear regression	Stagakis *et al.*, 2012
11	Pomegranate	Multispectral, thermal	Rotary	CSWI	Irrigation data	Direct comparison	Katsigiannis *et al.*, 2016

(pb0.001) for Ψ. The BGI1 index calculated from the blue (R400) and green (R550) bands resulted on the highest significance levels (pb0.001) for both Gs (r^2=0.62) and Ψ (r^2=0.49). Out of the structural indices assessed, RDVI, MTVI1 and TVI showed greater sensitivity for Gs (r^2=0.6; pb0.01) and Ψ (pb0.001) than NDVI. Chlorophyll fluorescence calculated from the micro hyperspectral imagery with the FLD3 method tracked stress levels, obtaining r^2=0.67 (pb0.05) with stomatal conductance, and r^2=0.66 (pb0.001) with water potential. The work presented in this manuscript demonstrates the feasibility of thermal, narrow-band indices and fluorescence retrievals obtained from a micro-hyperspectral imager and a light-weight thermal camera on board small UAV platforms for stress detection in a heterogeneous tree canopy where very high resolution is required.

Overall, the literature review on "**drones enabled assessment of moisture status of crops**" is summarized in Table 1. the data clearly demonstrated that drones and sensors usage in on-site moisture detection is possible that will facilitate rationalization of rare agricultural input "irrigation water". The review suggest the following:

1. Drones application in irrigation management is possible for agricultural and horticultural crops but major part of research focused on orchard crops, citrus, grapes and nectarines while field crops restricted to cotton and groundnut. There is lots of scope for expanding the technologies to other field crops particularly rice and maize where the water requirement is higher and extensive areas.
2. The moisture measurements using UAVs are mainly carried out using multispectral and thermal imaging and were exploited regardless of types of crops or agro-ecoregions
3. Rotary or fixed wing type of UAVs are useful for the assessment of moisture status of crops
4. In many case, drones enabled crop moisture estimation is performed using a standard set of variables namely NDVI, PRI, CSWI, WDI and GNDVI. Among them NDVI and CSWI are found to be universal and reported a robust indicator of moisture status.
5. The ground truth parameters to validated the moisture indices measured by drones include leaf water potential, stem water potential, leaf conductance, soil moisture status besides visual wilting score.
6. Statistical tool for establishing a robust correlation between aerial imageries and ground truth parameters is mainly linear regression

Overall, the data clearly demonstrate that drones with sensors can be effectively used to determine moisture status of the crops and rationalize irrigation water. More research is certainly needed to use the drone technology for rationalizing irrigation for small and marginal farms. The technology should be affordable and accessible to farmers for large scale adoption.

Chapter 7

Drones for Weed Management

Weeds are menace in agricultural production systems causing major reduction in yield as these set of plants compete for light, water and nutrients. Weeds can occur throughout fields and compete with the crop for resources negatively impacting on yields (Hamuda *et al.,* 2016). The yield reduction will be very severe in rainfed where there is a small window of moisture availability where both crops and weeds heavily compete for the resources. Chemical weed control is widely practiced due to the practical feasibility and quick response to the sprays. Depending on the infestation and type of cultivation form, chemical control can be used during pre- and post-emergence periods of crop growth. However, the use of herbicides can be costly and create environmental contamination, particularly if the chemical is applied uniformly across the field since the spatial variability of the weed infestation is not taken into account (Wang, Zhang, and Wei, 2019). Also, the possibility of increasing herbicide resistance in weeds can make it difficult and more expensive to manage. Thus, developing technologies to optimize the use of herbicides is required. This necessitated developing precision in weed management by developing sensors or other advanced tools and techniques that can be used in precision farming where drone technology may be linked for efficient and cost effective weed management systems.

7.1. Rice

Rice is one of the major crops of India being cultivated in various conditions such as puddle, semi-dry, aerobic and direct sown conditions. In all the cases, weeds are the prime matter of concern and several chemicals developed over the years and being used to control the weed menace in rice ecosystems. Rice belongs to grass family Graminacae and most of the wetland monocot weeds also belong to the same family or morphologically it resembles rice plants. This complexity makes selective weed control more complex and serious efforts are to be undertaken in order to differentiate grassy weeds with rice plants.

Barrero and Perdomo (2018) have developed a new method to fuse low resolution multispectral and high resolution RGB images is in order to detect Gramineae weed in rice fields with plants at 50 days after emergence (DAE). In their study, the images were taken from a fixed-wing unmanned aerial vehicle (UAV) at 60 and 70 m altitude. They evaluated a method that combines the texture information given by a high resolution red–green–blue (RGB) image and the reflectance information given by a low resolution multispectral (MS) image, to obtain a fused RGB-MS image with better weed discrimination features. After analyzing the normalized difference vegetation index (NDVI) and normalized green red difference index (NGRDI) for weed detection, it was found that NGRDI presents better features. The fusion method consists of decomposing the RGB image using the intensity, hue and saturation (IHS) transformation, then, a second order Haar wavelet transformation is applied to the intensity layer (I) and the NGRDI image. From this transformation, the low–low (LL) coefficients of the NGRDI image are replaced by the LL coefficients of the I layer. Finally, the fused image was obtained by transforming the new wavelet coefficients to RGB space. To test the method, a one hectare experimental plot with rice plants at 50 DAE with Gramineae weeds was selected. Additionally, to compare the performance of the method, two indices were used, specifically, the M/MGT index which is the percentage of detected weed area, and the MP index which indicates the precision of weed detection. These indices were evaluated in four validation zones using three Neural Networks (NN) detection systems based on three types of images; namely, RGB, RGB ? NGRDI, and fused RGB-NGRDI. **Barrero and Perdomo (2018)** have shown that the best weed detection performance in rice ecosystem was obtained by the NN with the fused image, with M/MGT index between 80 and 108 per cent and MP between 70 and 85 per cent.

7.1.1 Detection Performance of NN

In order to compare the detection performance of the NN using the fused image, three cases were studied. **First,** using only the texture descriptors from the RGB image; **second,** using texture descriptors from the RGB, and the average NGRDI descriptor from the NGRDI image; and **third,** using the texture and NGRDI descriptors form the fused image. To set up a valid comparison, the training and validation samples were taken at the same place for the three images, *i.e.*, RGB, NGRDI, and fused RGB-NGRDI. This was done with a synchronization tool of the free GNU license software Image (Image Processing and Analyzing in Java) (Schneider *et al.*, 2012). Tables 1 and 2 showed three examples of samples for each case used in training the NN.

Fig. 7.1. Examples of Negative and Positive Samples Used for Training the NN

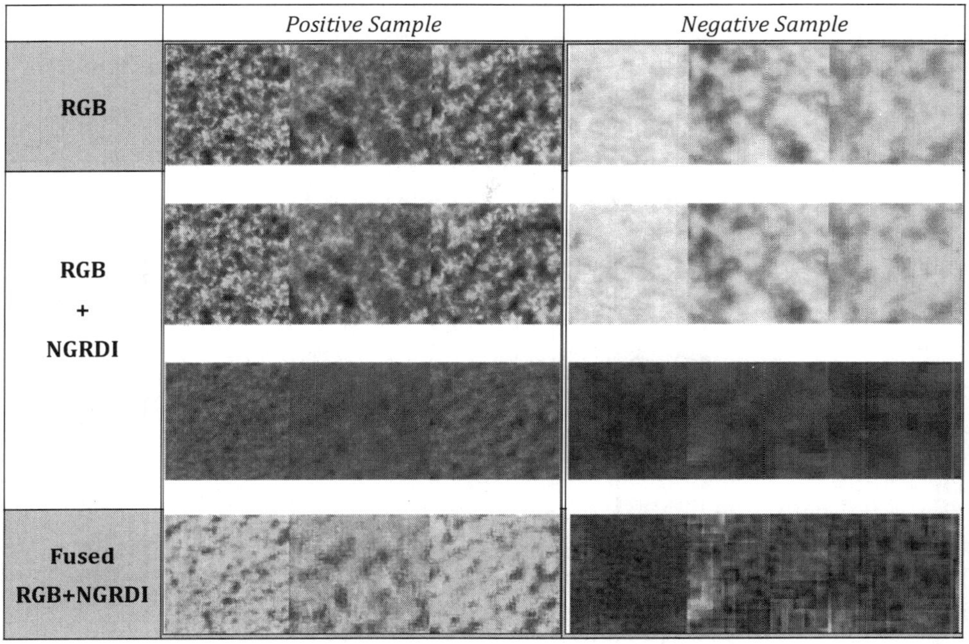

7.2.2 Performance Assessment

To assess the performance of the detection method quantitatively, three indices were evaluated, *i.e.*, first the **M index** which is the ratio of weed pixels to the total number of pixels of the image.

$$M(\text{per cent}) = \frac{\text{Image Number of Weed Pixels}}{\text{Total Image Number of Pixels}} \times 100 \text{ per cent}$$

Second, the **M/MGT index**, which is the ratio of the number of pixels that were detected as weed into the image in grayscale to the pixels that were selected as weed into the ground truth image. Hence, the ratio M/MGT indicates the percentage of pixels found to be weed.

$$M/MGT(\text{per cent}) = \frac{\text{Image Number of Weed Pixels}}{\text{Ground Truth Number of Weeds Pixels}} \times 100 \text{ per cent}$$

Third, the MP index, the ratio of the number of matching weed pixels between the ground truth image and the analyzed image to the total number of weed pixels of the image in grayscale was calculated. This index reveals the percentage of correct weed pixel detection. Additionally, to compute the **MP index**, a pixel matching algorithm was used. The algorithm compares the values of pixels at the same co-ordinate for the ground-truth and the images obtained from NN classification system.

$$MP(\text{per cent}) = \frac{\text{Number of matching Pixels}}{\text{Ground Truth Number of Weeds Pixels}} \times 100 \text{ per cent}$$

Fig. 7.2: Comparison of Ground-truth Image, in the Four Validation Zones, with the Results of the NN Classification System Based on RGB, RGB ? NGRDI, and Fused RGB-NGRDI Images

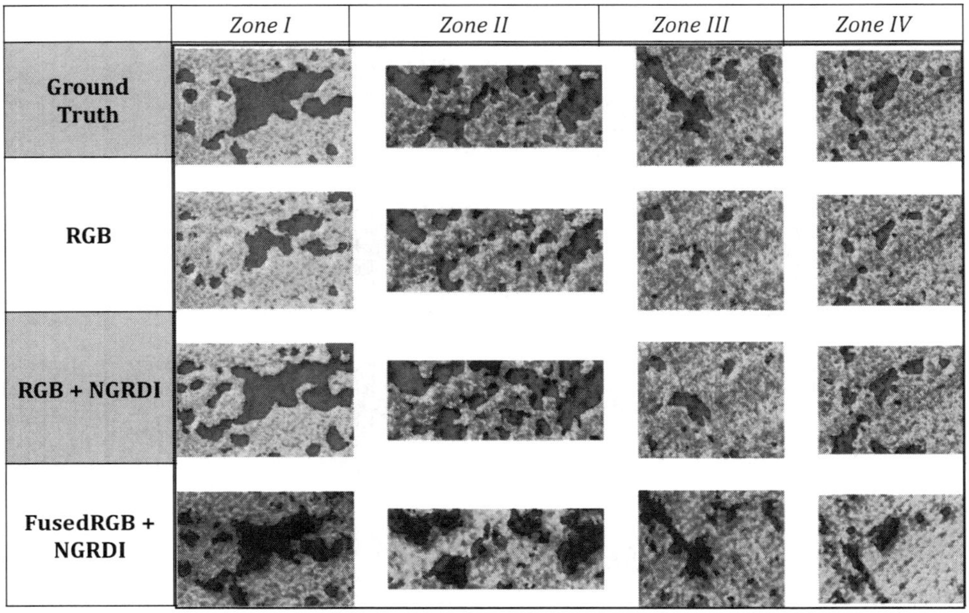

Barrero and Perdomo (2018) have shown the variations in **M/MGT** and MP indices for the three study cases in the four validation zones (Figure 6.1 a). For the case of using the RGB ? NGRDI images, it can clearly be seen that the NN over-detected weed plants because the values are over 100 per cent. Even though the case of using the fused RGB-NGRDI image were also over 100 per cent in two cases, the values were not far from 100 per cent, showing a better performance than the other two study cases. They have indicated **MP index** values for the three study cases and revealed that the best performance was shown by the fused RGB-NGRDI image with pixel matching between 70 and 85 per cent (Figure 6.1 b). Although the case of using RGB ? NGRDI images also presented high values, these were lower than the case of the fused image. Moreover, by analyzing the case of RGB ? NGRDI images for the two indices, it can be concluded that it is over-detecting

Overall, the study suggested that weed plants can be accurately detected by fusing the canopy reflectance of a field and a high resolution RGB image. Furthermore, images taken from a fixed-wing UAV at 60 and 70 m, from a 12.1 megapixels RGB and the RedEdge Micasense multispectral camera respectively, have enough information for detecting Gramineae weed in a rice field with plants

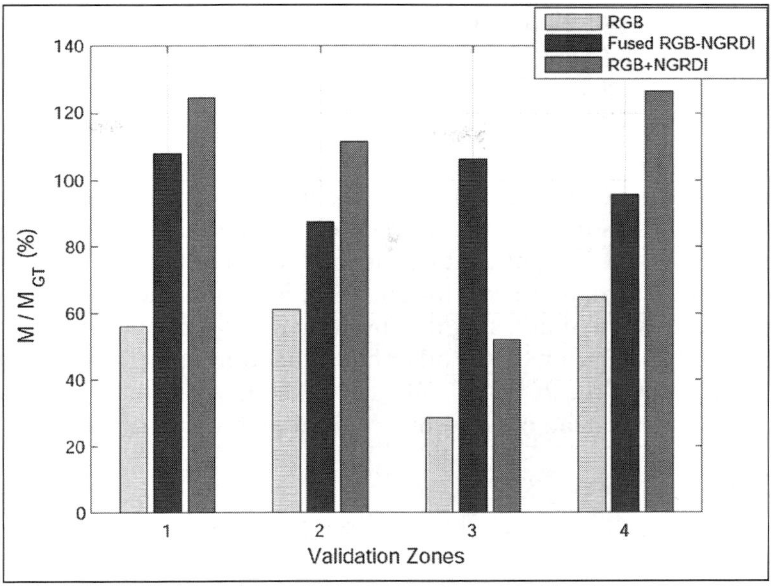

Fig. 7.3. M/MGT Comparison for the Study Three Cases in the Validation Zones.

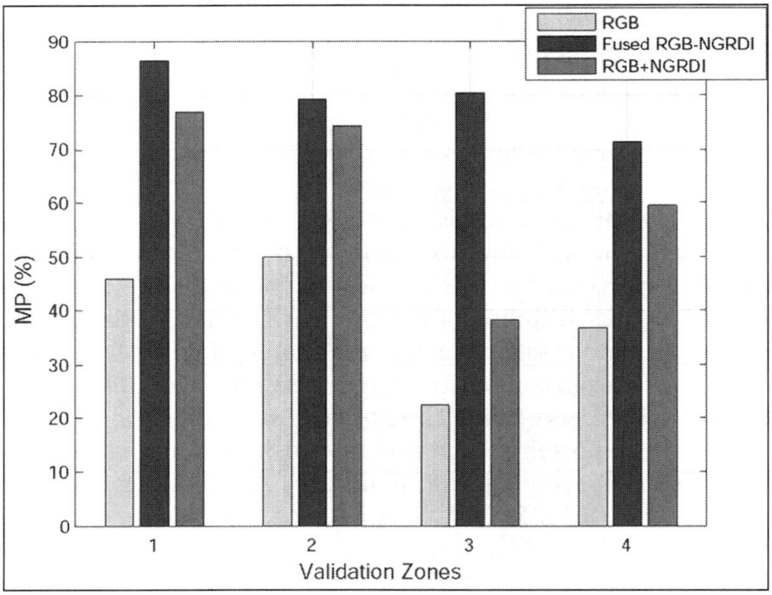

Figure 7.4. MP Index Comparison for the Three Study Cases in the Validation Zones.

at 50 DAE. Even though it can be considered that weed detection can be done just with canopy reflectance information of a multispectral camera, or with a non-professional high resolution RGB camera, or even using information from both images, this study has shown thatimage fusion with the method presented here improves the accuracy of weed detection. The results are supported by calculation of the M/MGT and MP indices, where the NN

detection system using the fused image outperformed the indices values for the NN detection systems based on RGB and RGB ? NGRDI images. The M/MGT index, which indicates the percentage of detected weed area, for the fused image were between 80 and 108 per cent on the four validation zones, indicating good detection performance; moreover, for the MP index which indicates the precision of the detection, the method using the fused image was the best with MP values between 70 and 85 per cent. On the other hand, the computing time using the fused image was half that of the RGB ? NGRDI image (4 h), because the NN system had to work on one image instead of two when calculating the descriptors.

7.2. Maize

Maize is cultivated extensively in India, a fast moving cereal in the country. The area under maize increased progressively over the years due to the reduction in area under cotton which is heavily infested by insect pests such as boll worms, stem weevil and sucking pests. Maize crop grows very vigorously at the early stage of the crop till 30-40 days and which coincides with threshold weed management. The selective weed control using mechanical device is the need of the hour and this requires basic study of discriminating weeds and maize plants. With this background, **P´erez-Ortiz *et al.* (2015)** have taken up extensive basic studies to discriminate the maize crop with weeds using object-based image analysis (OBIA) by means of supervised machine learning methods combined with pattern and feature selection techniques, devising a strategy for alleviating the user intervention in the system while not compromising the accuracy. They used a set of training patterns via clustering techniques so as to consider a representative set of the whole field data spectrum for the classification method. Furthermore, a feature selection method is used to obtain the best discriminating features from a set of several statistics and measures of different nature. Results from this research showed that the proposed method for pattern selection is suitable and leads to the construction of robust sets of data. The exploitation of different statistical, spatial and texture metrics represents a new avenue with huge potential for between and within crop-row weed mapping via UAV-imagery and shows good synergy when complemented with OBIA. Finally, there are some measures (especially those linked to vegetation indexes) that are of great influence for weed mapping in both sunflower and maize crops.

In another study, **Lopez Garnados *et al.* (2016)** have shown how to discriminate a highly competitive weed Johnsongrass in maize crops in many agricultural areas worldwide. To promote proper and efficient johnsongrass control according to current European regulations (as also suggested by other environmentally concerned authorities, farmers, herbicide distributers and consultants worldwide), the development of accurate tools to detect and map johnsongrass patches and to delineate the surface area requiring site-specific herbicide treatment based on the weed coverage is crucial. This study evaluated the performance of an OBIA algorithm on the imagery acquired with a UAV that was equipped with separate visible spectrum (RGB spectral range) and multispectral cameras (RGB+NIR spectral range) and that was programmed for three flight altitudes (30, 60 and 100 m) at Ribera and Espabera maize fields to classify and

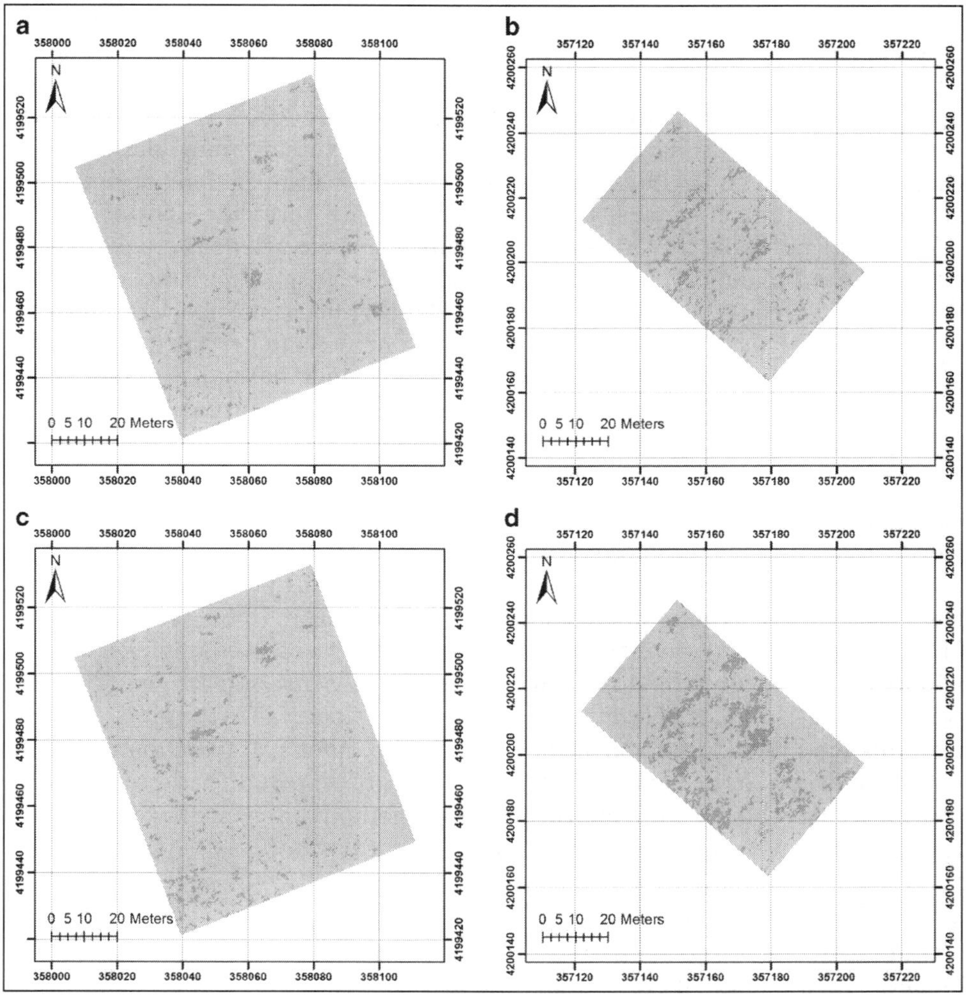

Figure 7.5: (a) Site-specific herbicide maps at Ribera; (b) Espabera maize fields using RGB camera and flying at 30 maltitude. Green indicates field surface requiring weed control. (c) Site-specific herbicide maps at Ribera; (d) Espabera maize fields using the multispectral (RGB+NIR) camera and 30 m altitude. Green indicates field surface requiring johnsongrass control.

map patches of johnsongrass (Figure 7.5). Then, as the most critical tool for the streamlining of herbicide application, a site-specific treatment map based on the most accurate weed coverage map was delineated for each location. For the overall accuracy of classification, the optimum results were found for the multispectral camera from a 30 m flight altitude, with slight differences between locations. The corresponding Johnsongrass coverage map was used for generating the site specific herbicide treatment maps that provided potential 96 and 85 per cent savings in herbicide use for the Ribera and Espabera maize fields. The identification of the areas that would not require treatment was certainly a relevant savings of very expensive herbicides, which could have further economic, agronomic and environmental repercussions. Although the accuracies for detecting Johnsongrass using the RGB camera at the same flight altitude were lower in both locations, we suggest that the farmer or the user of the UAV technology must decide which combination of flight altitude and camera will most address their objectives to ensure the detection of weeds and to minimize the herbicide use and to also consider the potential UAV battery limitations. Thus, the balance between sensor performance and improvements in accuracy must be weighed. In conclusion, with the use of UAVimagery and an automatic OBIA algorithm, satisfactory results were obtained for the accurate identification of johnsongrass patches and in the savings on herbicide use for controlling johnsongrass, one the most harmful and competitive weeds in maize. The results have strong implications because of the emphasis on opportunities to extrapolate these results to other areas worldwide, *e.g.* the Mediterranean basin, Argentina-Brazil and the corn-belt in the USA, among others.

Laursen *et al.* (2016) did a classic study to differentiate monocot and dicot weeds in relation to the maize crop. The aim of this study is to adapt the monocot and dicot coverage ratio vision (MoDiCoVi) algorithm to estimate dicotyledon leaf cover, perform grid spraying in real time, and present initial results in terms of potential herbicide savings in maize. The authors designed and executed an automated, large-scale field trial supported by the Armadillo autonomous tool carrier robot. The field trial consisted of 299 maize plots. Half of the plots (parcels) were planned with additional seeded weeds; the other half were planned with naturally occurring weeds. The in-situ evaluation showed that, compared to conventional broadcast spraying, the proposed method can reduce herbicide usage by 65 per cent without measurable loss in biological effect.

7.3. Sunflower

Sunflower is predominantly cultivated in rainfed system where weed menace is exorbitant. In order to overcome the weed infestation, selective weed control is necessary involving modern tools like drones. If weeds and crop can be discriminated by a sensing device, it will be of great help to selectively control weeds while minimizing the environmental hazard. In this context, **Lopez Granados *et al.* (2015)** have developed strategies for site-specific weed management which is defined as the application of customized control treatments only where weeds are located within the crop-field by using adequate herbicide according to weed emergence. They generated geo referenced weed seedling infestation maps in two sunflower fields by analyzing overlapping aerial images of the visible and near-infrared spectrum (using visible or multi-spectral cameras) collected by an unmanned aerial vehicle (UAV) flying at 30 and 60 m altitudes. The site-specific weed management program and weed mapping were done utilizing configuration and evaluation of the UAV and its sensors for image acquisition and ortho-mosaicking, as well as the development of an automatic and robust image analysis procedure. The control strategy was based on seven weed thresholds with 2.5 steps of increasing ratio from 0 per cent (herbicide must be applied just when there is presence or absence of weed) to 15 per cent (herbicide applied when weed coverage 15 per cent). As a first step of the imagery analysis, sunflower rows were correctly matched to the orthomosaicked imagery (Figure 7.6), which allowed accurate image analysis using object-based image analysis [object-based-image-analysis (OBIA) methods]. The OBIA algorithm developed for weed seedling mapping with ortho-mosaicked imagery successfully classified the sunflower-rows with 100 per cent accuracy in both fields for all flight altitudes and camera types, indicating the computational and analytical robustness of OBIA. Regarding weed discrimination, high accuracies were observed using the multi-spectral camera at any flight altitude, with the highest (approximately 100 per cent) being those recorded for the 15 per cent weed threshold, although satisfactory results from 2.5 to 5 per cent thresholds were also observed, with accuracies higher than 85 per cent for both field 1 and field 2. The lowest accuracies (ranging from 50 to 60 per cent) were achieved with the visible camera at all flight altitudes and 0 per cent weed threshold. Herbicide savings were relevant in both fields, although they were higher in field 2 due to less weed infestation. These herbicide savings varied according to the different scenarios studied. For example, in field 2 and at 30 m flight altitude and using the multi-spectral camera, a range of 23–3 per cent of the field (*i.e.*, 77 and 97 per cent of area) could be treated for 0–15 per cent weed thresholds. The OBIA procedure computed multiple data which permitted calculation of herbicide requirements for timely and site-specific post-emergence weed seedling management.

7.4. Sugarcane

De Souza *et al.* (2020) studied the spectral reflectance of weeds and sugarcane in order to develop site-specific application of herbicides which is highly desirable for optimizing its usage and reducing environmental damages. Thus, developing techniques for identification and mapping of weeds is necessary for a proper precision agriculture adoption. Such weed identification for site-specific management is difficult when the main crop is already established in the field. This study shows the possibility of differentiating sugarcane plants from weeds by the spectral behaviour of the leaves. The performance of two modeling methods, SIMCA (soft independent modelling by class analogy) and the RF algorithm (random forest) was compared. The simplification of the Vis-NIR spectrum into only four bands

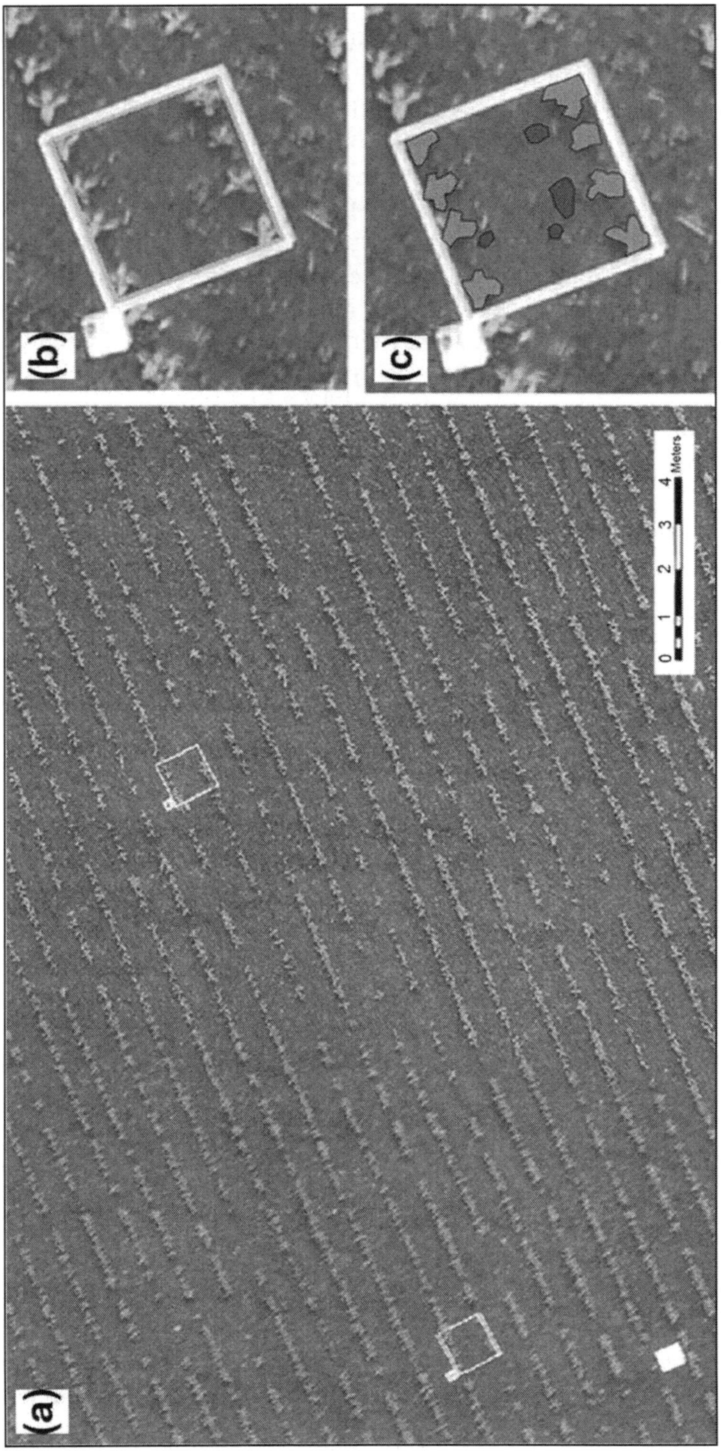

Figure 76: Partial View of the Ortho-mosaicked Imagery at 30 m Altitude (Sunflower field 2), Showing the Sunflower Rows, the Spectralon (White square placed between two sunflower rows at the bottom-left), some Patches of Weed Infestation and some of the 49 191 m Square Frames; (b) Detail of vector file created for every square frame (yellow); (c) Detail of the vector file created for the sunflower crop (green) and weed (violet) classes in one the 49 square frames.

of interest (500 - 550 nm; 650 - 750 nm; 1300 - 1450 nm; and 1800 - 1900 nm) was verified by demonstrating they had the same differentiation ability as the full visible-near infra-red spectrum. Thus, it was shown that performing the proper band selection and local calibration using a spectral classification approach may allow weed mapping and facilitate localized herbicide application.

De Souza *et al.* (2020) compared both models SIMCA and RA were compared to discriminate between weeds from the sugarcane crop, using Vis-NIR (visible-near infra-red) spectral response of the leaves. The classification models showed excellent performance (category A) according to the Kappa index (**SIMCA**: 0.9; **RF**: 0.96). This parameter indicates the agreement between the predicted and observed classes (Landis and Koch, 1977). The other evaluation measures (accuracy, sensitivity and selectivity) presented results with values greater than 80 per cent and the false positive rate reached a maximum of 1 per cent, confirming the good performance of the models. Obtaining a low false positive index is fundamental to ensure that no sugarcane leaf is confused with weed; otherwise, the herbicide application on the leaves of the crop could cause production damages. Good classification performance was achieved because the Vis-NIR spectra from the species showed different features (Figure 7.7). This indicates that there is variability in the constituent elements of each plant tissue, as well as their concentrations among the analysed species (Piron *et al.*, 2008; Shapira *et al.*, 2013).

Figure 7.7 clearly indicate that there is a unique and distinctive vis-NIR pattern which can be used to develop sensors that can be fitted in drones for précised and selective weed control.

7.5. Lettuce and Broccoli

Gai *et al.* (2019) explored the possibility of use of robotics for weeding. Robotic weeding enables weed control near or within crop rows automatically, precisely and effectively. A computer vision system was developed for detecting crop plants at different growth stages for robotic weed control (Figure 7.7). Fusion of color images and depth images was investigated as a means of enhancing the detection accuracy of crop plants under conditions of high weed population. In field images of broccoli and lettuce were acquired 3–27 days after transplanting with a Kinect v2 sensor. The image processing pipeline included data preprocessing, vegetation pixel segmentation, plant extraction, feature extraction, feature based localization refinement, and crop plant classification. For the detection of broccoli and lettuce, the color depth fusion algorithm produced high true positive detection rates (91.7 per cent and 90.8 per cent, respectively) and low average false discovery rates (1.1 per cent and 4.0 per cent, respectively). Mean absolute localization errors of the crop plant stems were 26.8 and 7.4mm for broccoli and lettuce, respectively. The fusion of color and depth was proved beneficial to the segmentation of crop plants from background, which improved the average segmentation success rates from 87.2 per cent (depth based) and 76.4 per cent (color based) to 96.6 per cent for broccoli, and from 74.2 per cent (depth based) and 81.2 per cent (color based)

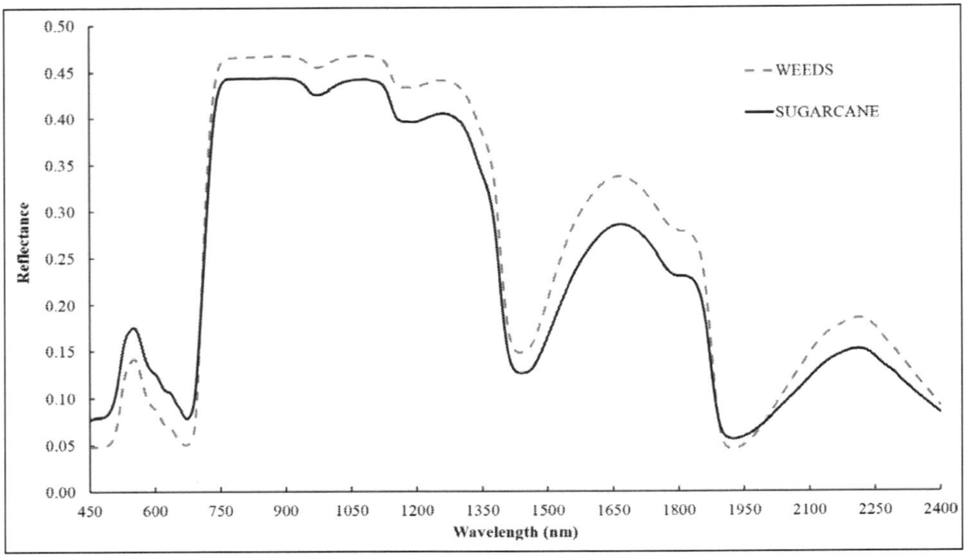

Figure 7.7: Mean Vis-NIR Spectra from Sugarcane and Weeds Leaves.

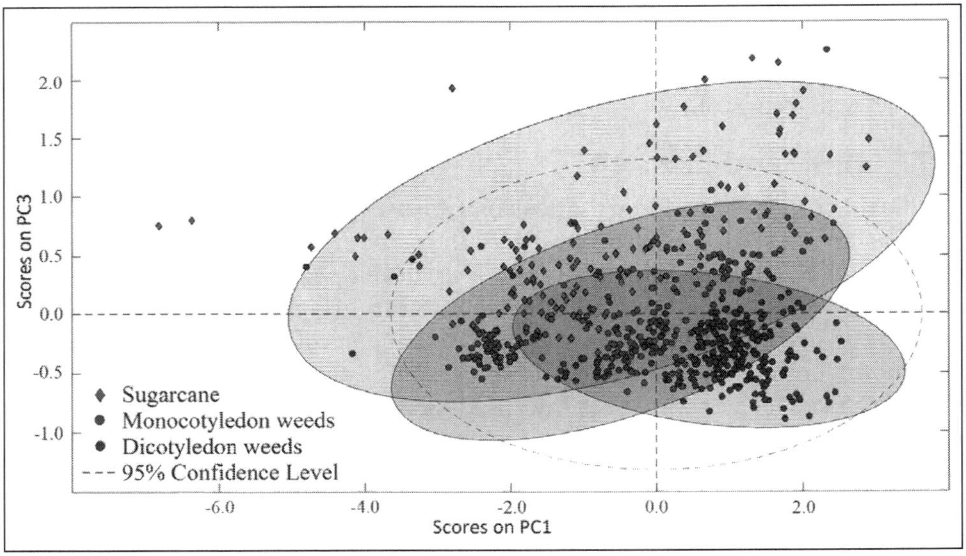

Fig. 7.8: Principal Component Analysis (Scores) of the Vis-NIR Spectra Coloured According to Plant Class. Ellipses indicate the limit of each class.

to 92.4 per cent for lettuce, respectively. The fusion based algorithm had reduced performance in detecting crop plants at early growth stages.

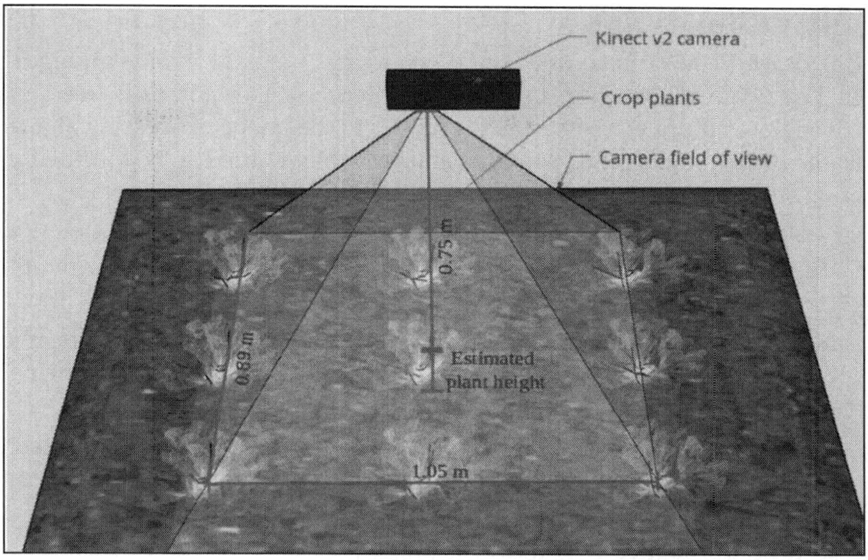

Figure 7.9: The Kinect sensor was mounted at 0.75m plus the estimated plant height above the field surface and had a1.05m× 0.89m field of view of the plants and soil below [Color figure can be viewed at wileyonlinelibrary.com].

Figure 7.10: Remote controlled data collection apparatus in the horticulture research station of Iowa State University [Color figure can be viewed at wileyonlinelibrary.com].

7.6. Algorithm Design

Gai *et al.* (2019) have developed sequential steps to evolve an algorithm design. The acquired depth images from the depth sensor and RGB color images from the color camera were used as the inputs to the image processing algorithm. The framework of the detection and localization algorithm can be outlined in the following steps:

- ☆ **Step 1. Preprocessing:** This procedure removed invalid pixels and noise pixels in point clouds. A useable area filter, a cutoff filter, and a simplified neighbor count filter were used.
- ☆ **Step 2. Segmentation:** The background was removed by detecting the soil surface (assumed to be a plane) using both color and depth information. Vegetation pixels were extracted.
- ☆ **Step 3. Plant extraction**: The vegetation pixels were separated into different clusters using their spatial relationship. Each cluster represented one individual plant. These plants were localized as well.
- ☆ **Step 4. Feature extraction**: Canopy and leaf features of each detected plant were extracted.
- ☆ **Step 5. Feature based localization refinement**: The extracted features were used for refining the localization and extraction results.
- ☆ **Step 6. Classification:** Crop/weed classification was applied to all the separated plants based on their extracted features using machine learning techniques.

Gai *et al.* (2019) proposed an image processing pipeline for detecting and localizing broccoli and lettuce crop plants by fusing color and depth images. For both crops (broccoli and lettuce) the detection algorithms produced high true positive detection rates (91.7 per cent and 90.8 per cent, respectively) and low average false discovery rates (1.1 per cent and 4.0 per cent, respectively). The average localization errors of the crop plant stems were 26.8 and 7.4mm for broccoli and lettuce, respectively. The fusion of color and depth improved the average crop plant segmentation success rates from 87.2 per cent (depth based) and 76.4 per cent (color based) to 96.6 per cent for broccoli, and from 74.2 per cent (depth based) and 81.2 per cent (color based) to 92.4 per cent for lettuce, respectively.

The fusion based algorithm had reduced performance in detecting crop plants at early growth stages. From the result of the experiments conducted, they concluded the following:

- ☆ The fusion of color and depth is beneficial to the segmentation of plants from background with same algorithm complexity at most tested growth stages. However, the fusion is less effective than using color only at early growth stages (broccoli DAT < 7; lettuce DAT < 12).

Table 3. Drones in Selective Weed Control in Agricultural and Horticultural Cropping Systems

Crops	*Sensor*	*UAV Type*	*Model Input*	*Model Output*	*Model Type*	*Weeds/Comments*	*Reference*
Rice	Multispectral RGB	UAV	NDVI and NGRDI	Fused RGB-NGRDI images	Neural Network design	Detecting Gramineae weed	Barrero and Perdomo (2018)
Maize	Metal–oxide–semiconductor (CMOS) sensor.	UAV	Pixel	Potential herbicide savings	Monocot and dicot coverage ratio vision (MoDiCoVi) algorithm	Estimate the ratio of dicotyledon pixels to monocot pixels	Laursen *et al.* (2016)
Maize	Multispectral sensor	UAV	NDVI	Mapping of Johnson grass	OBIA algorithm	*Sorghum halepense* (johnsongrass)	Lopez Garnados *et al.* (2016)
Maize	Still camera	UAV	Super-pixels	Weed mapping	Object-Based Image Analysis (OBIA)	Pigweed (*Amaranthus blitoides*	Perez-Ortiz *et al.* (2015)
Barley	RGB	Rotary	Excess green index	Weed harrowing impact	Linear regression	-	Rasmussen, *et al.*, 2013
Sorghum	RGB	Rotary, fixed wing	Excess green index	Weed infestation	K-means	Palmer amaranth, barnyardgrass, Texas *Panicum, morningglory*	Shi *et al.*, 2016
Soybean	Hyperspectral	Rotary, fixed wing	**UAV** Images	Weed detection	Visual inspection	Palmer amaranth, Italian ryegrass	Huang, *et al.*, 2018
Sugarcane	RGB	Rotary	Statistical image descriptors	Crop and weed classification	Random forest	*Tridax daisy*, sourgrass	Yano *et al.*, 2016
Lettuce	RGB	-	Vegetation pixel	Crop plant classification	Fusion algorithm	-	Gai *et al.* (2019)

Crops	*Sensor*	*UAV Type*	*Model Input*	*Model Output*	*Model Type*	*Weeds/Comments*	*Reference*
Broccoli	RGB	-	Vegetation pixel	Crop plant classification	Fusion algorithm	-	Gai *et al.* (2019)
Grapes	Multispectral, thermal	Rotary UAV	Thermal data, NDVI	Heat stress quantification	Linear regression	-	Gennaro, *et al.*, 2017
Moss	Hyperspectral	Rotary UAV	Spectral bands	Chlorophyll content, leaf density	Support vetor regression	-	Malenovsky, *et al.*, 2017

- Enhanced by the fusion of both color images and depth images, the crop plant detection algorithm was capable of detecting and localizing broccoli and lettuce plants at different growth stages with high weed densities in the background of the images.

The framework of the proposed algorithm can be extended to different crop species, and the developed algorithm can be applicable to other applications such as selective spraying and plant mapping.

Overall, the use of drones in weed management has been extensively studied and the available data are summarized in **Table 2.**

- Selective herbicide application using sensors and drones have been attempted in rice, maize, wheat, sugarcane, grapes and vegetables and proved effective for further studies and commercial use of the technologies
- The UAVs utilized for weed management primarily exploit RGB or multispectral or still cameras for detection, discrimination and develop strategies for site specific herbicide application
- The UAVs meant for effective weed control includes ordinary or rotary wing type to undertake this task in precision farming
- Major indices used for weed diversity and species specificity include NDVI, Excess Green Index, Vegetation Index, RGB *etc.* These indices found to be useful in a wide array of cropping systems.
- The output from the drone technology encompasses diversity, species specificity; weed mapping, discriminating monocots with dicots and weed control efficiency *etc.*
- Computer aided models that are used for selective weed management neural network, OBIA (Object Based Image Analysis), linear regression and fusion algorithms
- Drone technology in weed management is fairly recent and the review suggests huge potentials and more research to be done to commercialize the technology for large scale adoption in smart farming.

Chapter 8

Drones for Pests and Diseases Management

The pests and diseases are the prime factors causing extensive crop damage and food losses around the world. The Food and Agriculture Organization of the United Nations has predicted that the losses due to pests and diseases have been estimated over 37 per cent (Cao, 2015). They also severely affecting the crop growth, yield and quality of produce (Gossen *et al.*, 2008; Berger Neto *et al.*, 2017). Recently, the crops cultivated in India being threatened by invasive pests like fall army worm (*Spodoptera frugiperda* JE Smith) in maize and Rugose spiraling whitefly in coconut (*Aleurodicus rugiperculatous* Martin) causing extensive damage during the year 2018 and 2019 (Lal and Bikram, 2019). These pests are to be meticulously monitored and proper technology capsule to be adopted to save the crops from devastation. Several plant protection strategies are being followed in an integrated way to ensure that crops are protected during the entire crop growing season.

The plant protection measures are to be taken as community basis so as to ensure effective management of pests and diseases. In India, more than 80 per cent of the farm lands are in the category of small and marginal (< 1 ha), it is very difficult to manage the invasive pests. If one field is sprayed, the pests simply shift their feeding to the neighboring fields. In order to cope with high-risk pests and diseases, the annual global use of plant protection products in pests and diseases was more than 3 billion kg (Heidary *et al.*, 2014). The effective utilization pesticides that are sprayed on the crops hardly exceeds 20-30 per cent and the remaining 70-80 per cent goes as run-off, leaching, evaporation and drift that cause soil and aquatic pollution besides deteriorating the quality of the crop produce (Markle *et al.*, 2016; Torrent *et al.*, 2017).

Under these circumstances, effective and timely spray of plant protection measures are very important for which UAV shows a great potential in the field of agricultural aviation with the advantages of high efficiency, labor force and resources saving, strong adaptability (suitable for coping with sudden disasters) and environmental protection (Meng *et al.,* 2018; Shamshri *et al.,* 2018). The agricultural research institutions State Department of Agriculture have bestowed attention to design and fabricate drones that suit Indian conditions.

The design of UAVs should consider various parameters such as droplet size, wind speed, flight speed and flight height (Zhang *et al.,* 2012; Qin *et al.,* 2014; Zhang *et al.,* 2015). Further, meteorological parameters like wind speed, temperature and relative humidity can affect the efficacy of pesticide sprays under field conditions (Wang *et al.,* 2018). Under natural conditions, it is very difficult to control the meteorological parameters and thus scientists have attempted to study the drones under protected conditions.

Lv *et al.* (2019) have examined the efficacy of drones under the protected environment without the interference from external factors, the influence of UAV flight speed on droplet deposition and distribution uniformity. The study is very useful in optimizing various parameters to determine the effective spray of pesticides using drones. One of the most important criteria for the successful delivery of pesticides or any other inputs is droplet size. Lv *et al.* (2019) conducted a classic study to evolve a droplet deposition index.

Droplet Deposition Indices

The droplet deposition indices are employed to evaluate the droplet deposition effect accurately, mainly including droplet deposition density, droplet deposition coverage, and arithmetic mean of droplet size and variation coefficient (Zhu *et al.,* 2011). Among them, ***droplet deposition density*** refers to the number of droplets deposited on per unit area of droplet collection material, which was water-sensitive paper in this study. ***Droplet deposition coverage*** refers to the area of all droplet particles per unit area on the droplet collection material (Cunha *et al.,* 2012). ***Arithmetic mean of droplet size*** refers to the average value of all droplet diameters in one spray sample (Fan *et al.,* 2011). ***Variation coefficient CV*** refers to the distribution uniformity of droplet deposition in aerial spraying operation.

Those four evaluation indexes could be calculated

a. **Droplet deposition density** (D) = n/A

   n is the number of droplets deposited on the collection material

   A is the area of droplet collection material.

b. **Droplet deposition coverage** (C) = S/A

   S is the area of droplet deposition particle

   A is the area of droplet collection material

A

B

C

D

E

F

(a) Fixed Wing (Department of Remote Sensing and GIS, TNAU, Coimbatore); (b) Helicopter (Hornet Maxi); (c) Quad Copter (Department of Remote Sensing and GIS, TNAU, Coimbatore); (d) Blimp; (e) Flapping-wing (SmartBird); and (f) Parafoil-wing (Tetracam). (Pg 16)

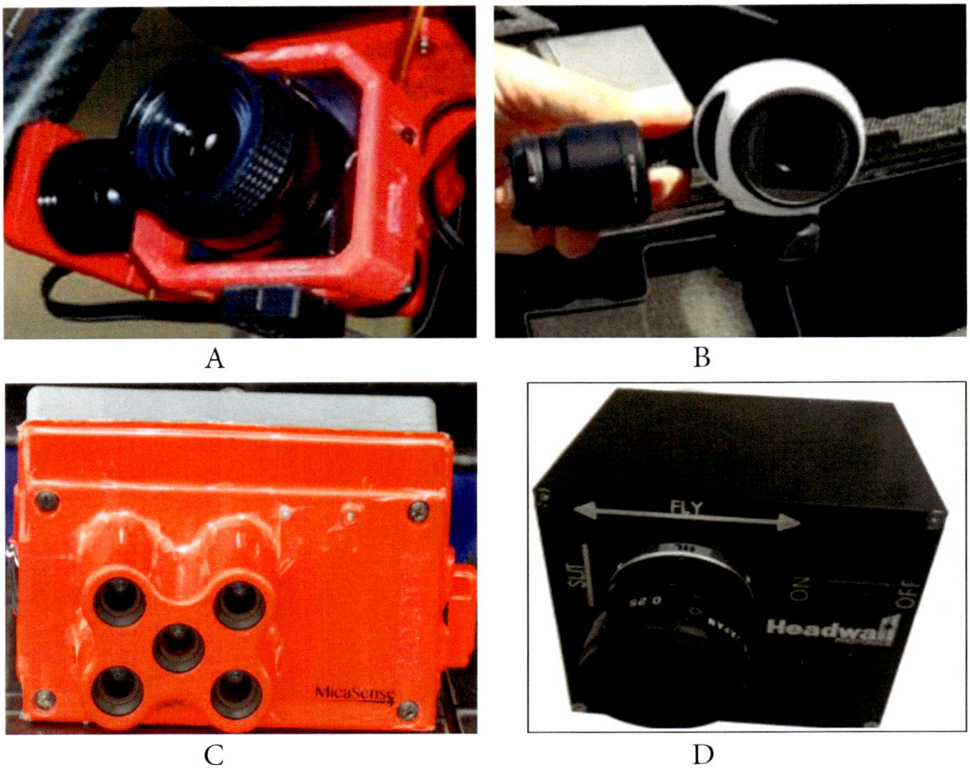

Examples of Sensors Used by UAVs for PA: (a) Thermal sensor (Maimaitijiang *et al.,* 2017); (b) RGB sensor (Hassanein *et al.,* 2018); (c) Multi-spectral sensor (Department of Remote Sensing and GIS, TNAU, Coimbatore); and (d) Hyperspectral sensor (Bhandari *et al.,* 2018). (Pg. 23)

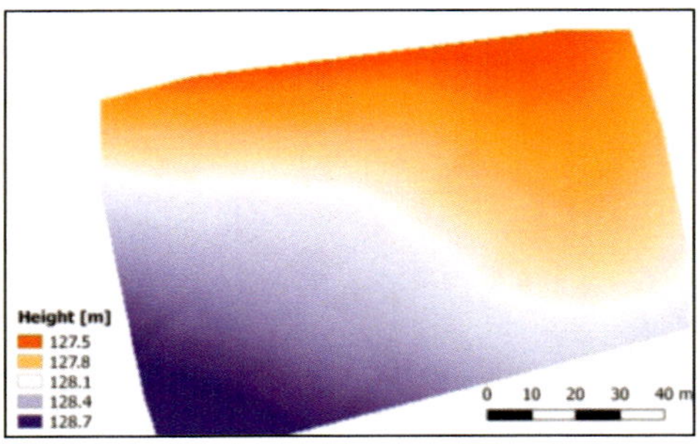

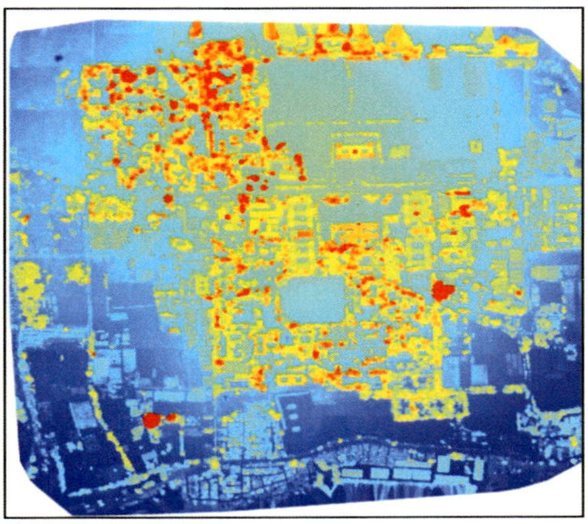

(a) Digital Terrain Model (b) Digital Surface Model of TNAU, Coimbatore (Pg 27)

NDVI (Normalized Difference Vegetation Index) of the cotton field (TNAU, Coimbatore). (Pg 32)

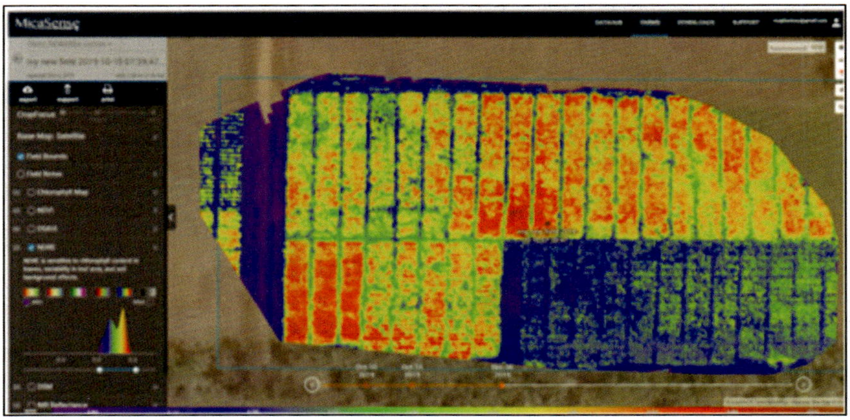

(b) NDRE (Normalized Difference Red-edge index) of the cotton field (TNAU, Coimbatore). (Pg 32)

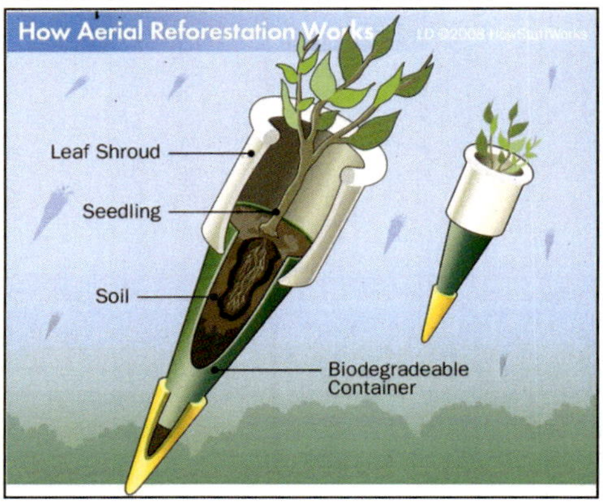

Page 37

Multi-level crop phenotype monitoring platform; (b) UAV crop phenotype monitoring platform; (c) IoT phenotype monitoring platform; (d) Mobile vehicle-mounted monitoring platform (Source: Shi *et al.*, 2020) (Pg. 47)

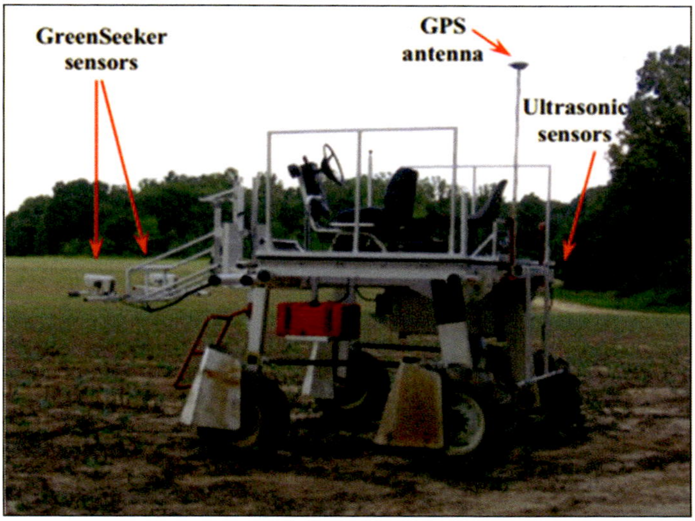

Variable Fertilizer Applicator for Cotton (Page 48)

The six-rotor UAV with a domestic digital camera; (b) the ground monitor displaying real-time imagery and other data transmitted from the UAV. (Page 55)

a Back view of the sprayer. **b** Closer view of the lighting chamber.

c Inside of the lighting chamber: front view. **d** Inside of the lighting chamber: side view.

Variable-rate Fertilizer Applicator for Corn (Page 49)

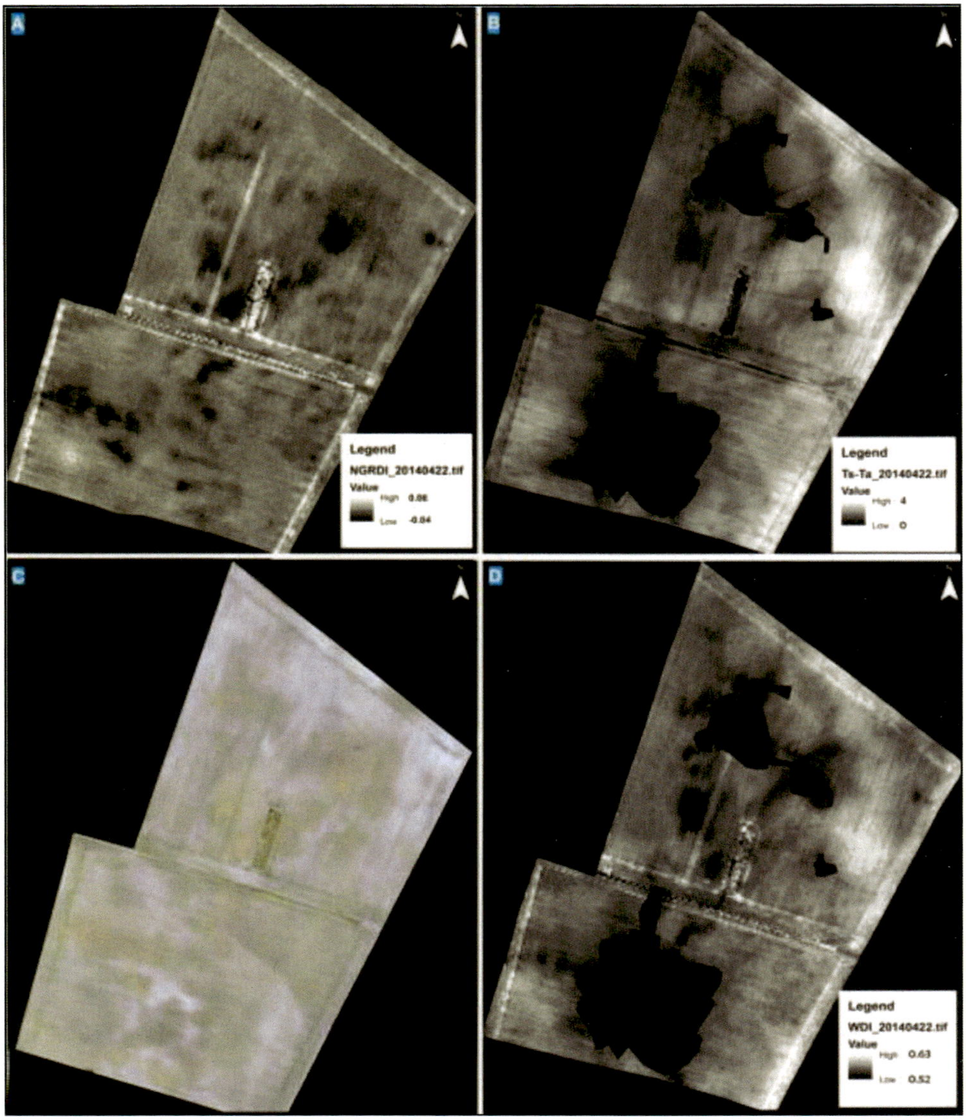

Imagery and Indices from April 22. (a) NGRDI; (b) TS-TA; (c) Orthophoto (RGB data) and D) WDI. Black areas (no data) within the field on maps (b) and (d) are due to insufficient texture in images collected in these areas, which meant that the images could not be aligned with the Structure-from-Motion algorithm. (*Source*: Hoffmann *et al.*, 2016). (Pg 65)

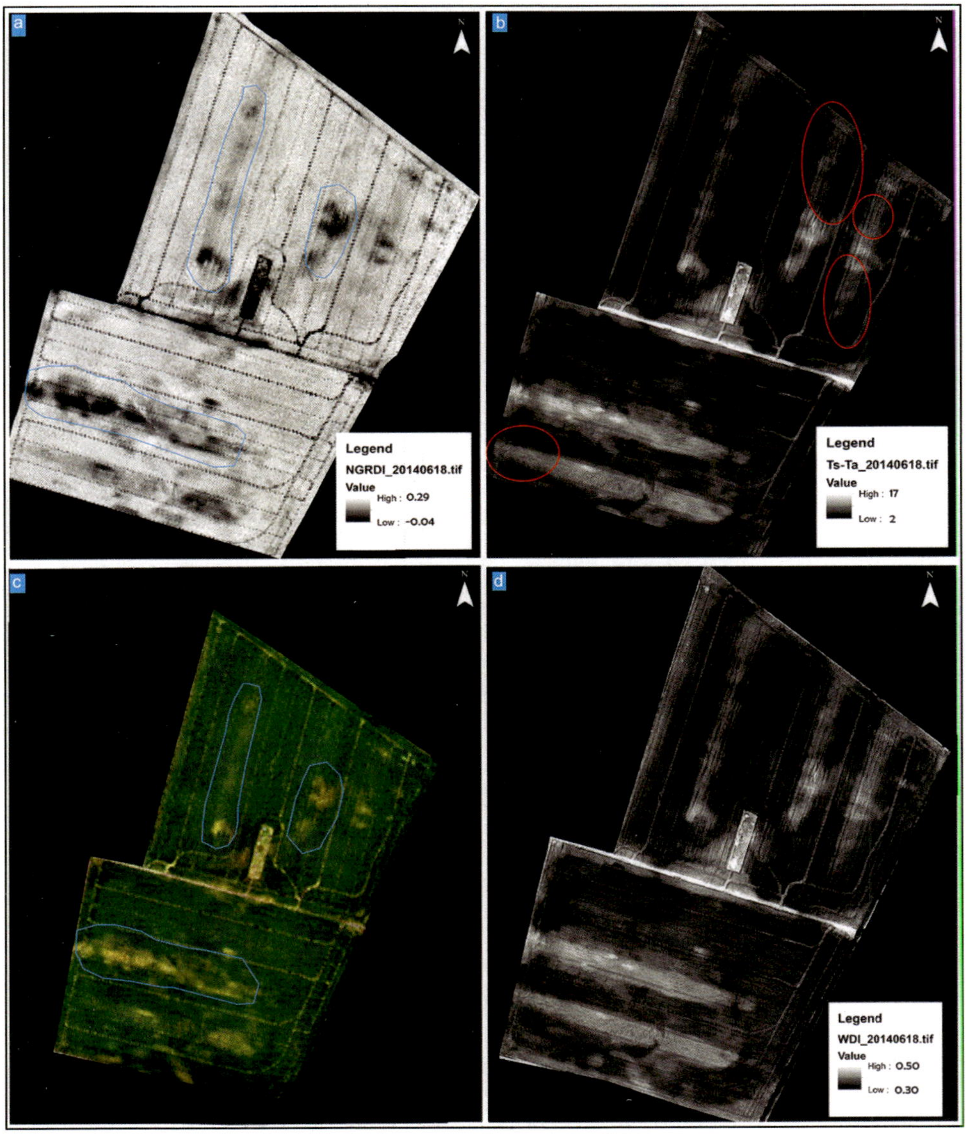

Imagery and Indices from 18 June. (a) NGRDI; (b) TS - TA; (c) Orthophoto (RGB data) and (d) WDI. Red circles in (b) are explained in Sect. 4.2. Areas highlighted with blue are examples of locations with ripe crops (*Source*: Hoffmann *et al.*, 2016). (Pg 66)

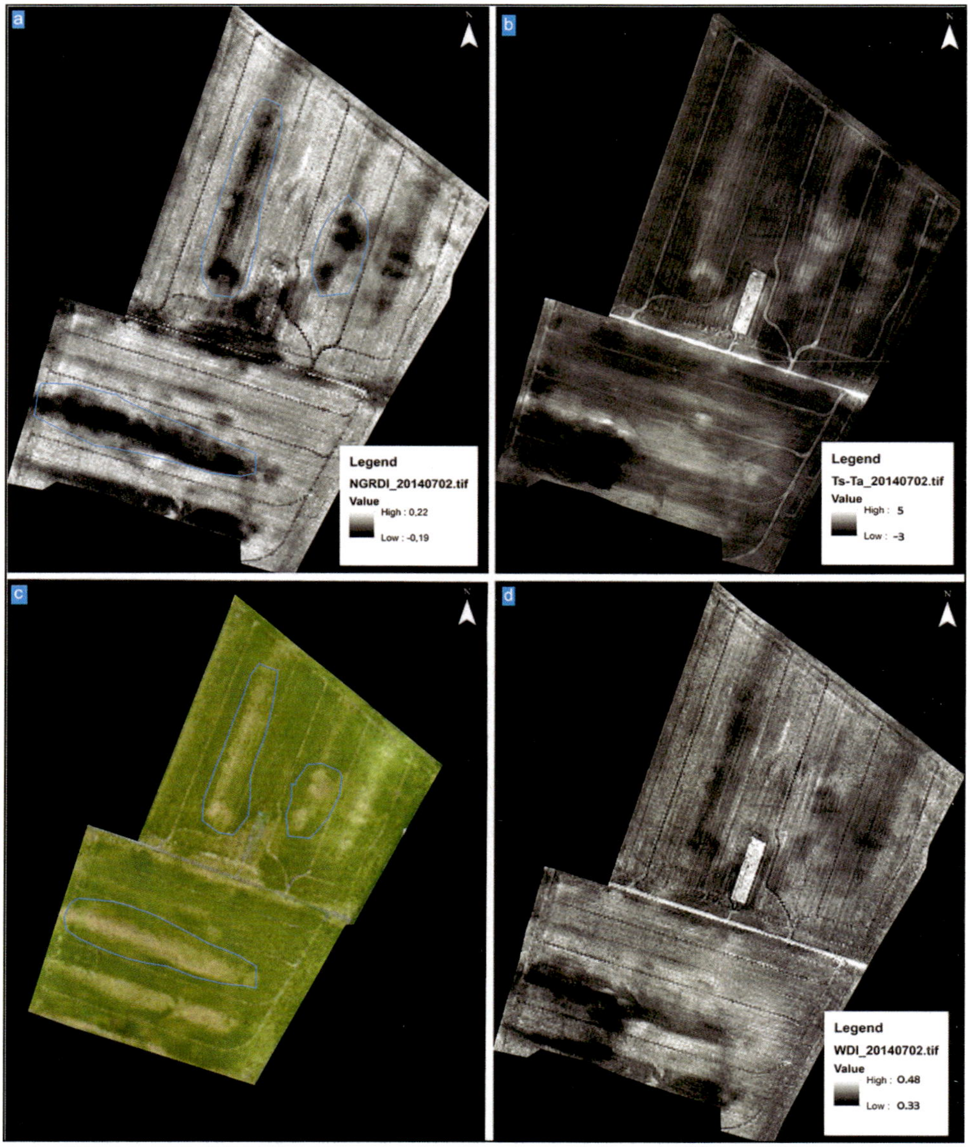

Imagery and Indices from 2 July. (a) NGRDI; (b) TS - TA; (c) Orthophoto (RGB data) and (d) WDI. Areas highlighted with blue are examples of locations with ripe crops (*Source*: Hoffmann *et al.,* 2016). (Pg 67)

Aerial view of the orchard and its irrigation management units. Al, Or, Ap, Le and Pe correspond to almond, orange, apricot, lemon and peach, respectively. Solid line indicate the plots that were fully irrigated (-ww plots), while dotted lines correspond with the plots where irrigation was withheld prior to the flights(-d plots). (Pg 74)

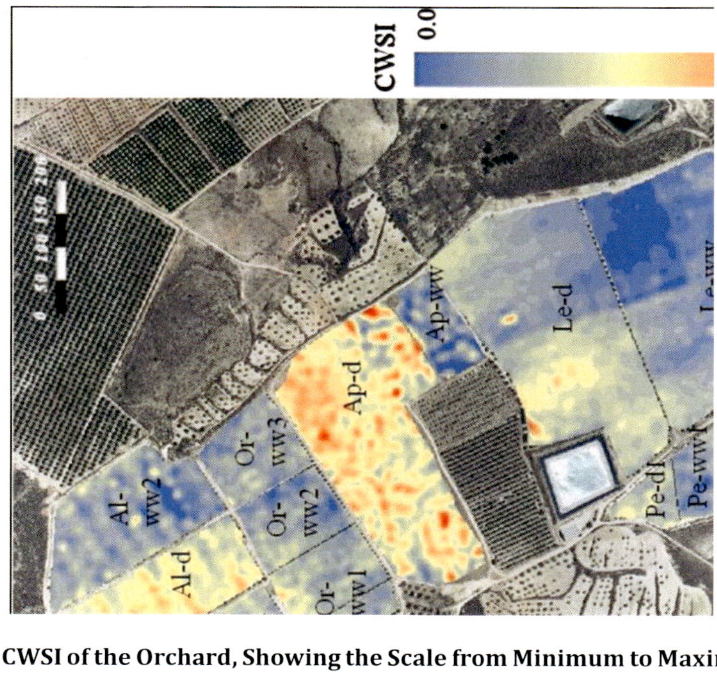

Map of the CWSI of the Orchard, Showing the Scale from Minimum to Maximum CWSI (*Source*: Gonzalez-Dugo *et al.*, 2013). (Pg 75)

Examples of Negative and Positive Samples Used for Training the NN (Pg 83)

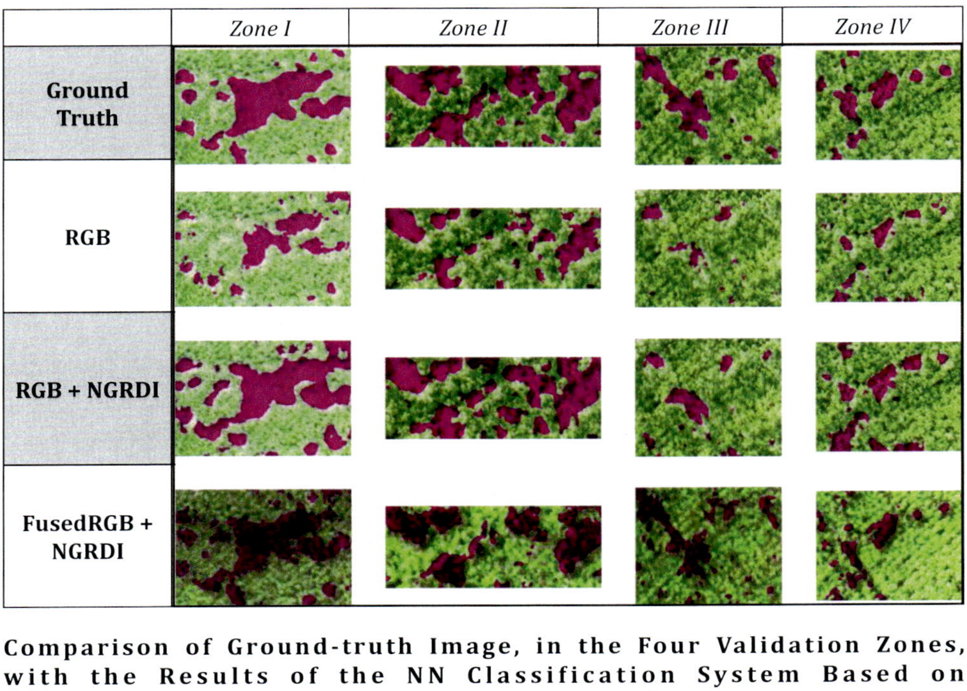

Comparison of Ground-truth Image, in the Four Validation Zones, with the Results of the NN Classification System Based on RGB, RGB ? NGRDI, and Fused RGB-NGRDI Images (Pg 84)

Partial View of the Ortho-mosaicked Imagery at 30 m Altitude (Sunflower field 2), Showing the Sunflower Rows, the Spectralon (White square placed between two sunflower rows at the bottom-left), some Patches of Weed Infestation and some of the 49 191 m Square Frames; (b) Detail of vector file created for every square frame (yellow); (c) Detail of the vector file created for the sunflower crop (green) and weed (violet) classes in one the 49 square frames. (Pg 90)

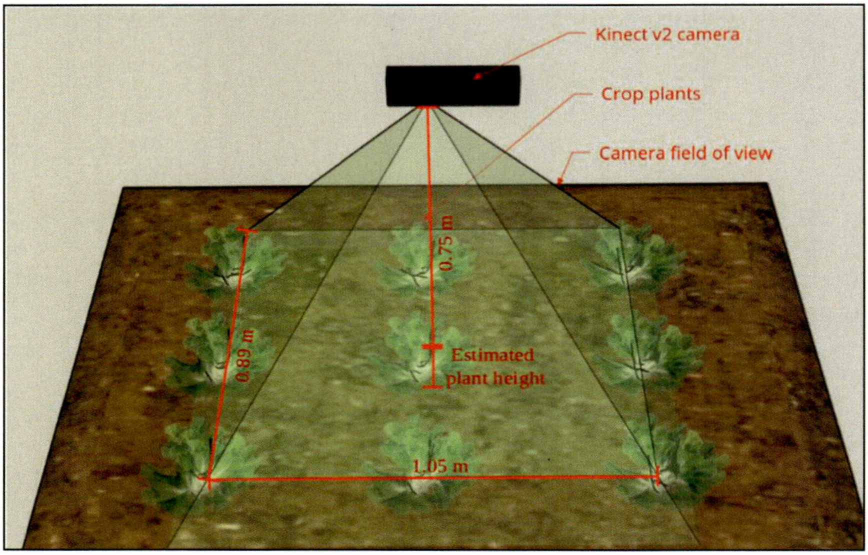

The Kinect sensor was mounted at 0.75m plus the estimated plant height above the field surface and had a1.05m× 0.89m field of view of the plants and soil below [Color figure can be viewed at wileyonlinelibrary.com].(Pg 93)

Remote controlled data collection apparatus in the horticulture research station of Iowa State University [Color figure can be viewed at wileyonlinelibrary.com]. (Pg 93)

c. **Arithmetic mean of droplet size** (D_0) = Σ Di Ni/Σ Ni

Di is the droplet diameter of a certain size interval

Ni is the number of droplets at a certain interval

d. **Variation coefficient CV** = SD/X

$$\text{Standard Deviation (SD)} = \sqrt{\frac{\sum_{i=1}^{n}(x_i - \mu)^2}{n-1}}$$

Xi is the droplet deposition particle size per unit area of each sample card

μ is the average droplet size per unit area of each sample card

SD is the standard deviation

n is the number of droplets per sample card.

The smaller the variation coefficient *CV*, the more uniform the droplet distribution in the target spraying area, and the better the pesticide effect plays. X

Infrared Thermal Imaging

Lv *et al.* (2019) have used the infrared thermal imaging technique to accurately measure the droplet deposition. After the drones spray, there will be changes in leaf temperature and the infrared thermal imaging technique which is able to detect infrared specific band signal of object thermal radiation by optoelectronic technology, which was used as a supplementary means for droplet deposition measurement. The experiment can be done in a closed environment and the thermal image acquisition was conducted immediately after the spray test. In order to avoid interference at the beginning and ending of the spray process, the number 2, 3 and 4 plants in the middle target area can be selected as the temperature measurement objects of the infrared thermal imager. And the flight speed was set at 0.5 m/s, 0.7 m/s and 0.9 m/s respectively. A classic method developed by Lv *et al.* (2019) can be employed to determine the droplet deposition by the drones.

Studies of Lv *et al.* (2019) have concluded that (i) the deposition density, droplet deposition coverage and arithmetic mean of droplet size *D0* decreases as the flight speed increases, (ii) the droplet diameter variation coefficient *CV*, which reflects the distribution uniformity of droplet size, increases slowly with the acceleration of flight speed. That is, the distribution uniformity of particle size decreases with the increase of flight speed, indicating that the increase of the UAV flight speed results in the worse droplet distribution uniformity and (iii) the temperature change rate *R* shows a linear decline trend with the acceleration of flight speed. It is highly consistent with the variation trend of droplet deposition coverage and arithmetic mean of droplet size *D0*, demonstrating the practicability of infrared thermal imager in evaluating the droplet deposition in the field of aerial spraying.

8.1. Rice

8.1.1 Aerial Spray for Brown Plant Hoppers in Rice

Rice is a staple food crop of more than 2.7 billion people in Asia and the loss of productivity of the crop has been estimated as more than 20 per cent.

The brown plant hopper (BPH) *Nilaparvata lugens* is a typical sap-sucking insect of rice and has caused considerable crop loss globally, especially in southern China and major part of India in the late season rice crop planted during September - October. The BPH causes damage at the late stage of rice growth. During the late stage of the crop, it is very difficult for undertaking manual spray as the leaves of the rice canopy overlap. Further, BPH often colonized at the lower part of the plant which is inaccessible through a manual sprayer. In addition, the muddy fields and overlapping plants make the conventional system of pesticide spray becomes hardly possible. In order to overcome the bundle of practical difficulties and acute labour shortage, aerial spray of pesticide using UAVs become inevitable.

Qin *et al.* (2016) have explored a possibility of using a small unmanned aerial vehicle (UAV) that can spray pesticide with high efficiency and with no damage to the rice crop being cultivated in small fields in hilly mountains of China. They studied the influence of spraying parameters, such as operation height and operation velocity, of the UAV on droplet deposition on the rice canopy and protection efficacy against plant hoppers. In their experimental set up, they sprayed chlorpyrifos, droplets of 480 g L^{-1} (Regent EC) (at a dose of 432 g a.i. ha^{-1}, spray volume rate of approximately 15 L ha^{-1}). They found that the deposition and distribution of droplets in the late stage of rice growth was closely related to the operational height and velocity of crop spraying as executed by the UAV, further affecting insect control. When the spraying height was 1.5 m and the spraying velocity 5 m s^{-1}, the droplet deposition in the lower layer was maximized, and the droplets exhibited the most uniform distribution (CV = 23 per cent). The insecticidal efficacy was 92 per cent to 74 per cent from 3 to 10 days after spraying insecticide. Both the insecticidal efficacy and the persistence period were greater than those achieved with a hand lance operated from a stretcher-mounted sprayer (at dose of 432 g a.i. ha^{-1}, spray volume rate of approximately 750 L ha^{-1}), especially on the 5th day, indicating that UAV had a low volume and highly concentrated spray pattern to enhance the duration of efficacy. This work offers a basis for the optimized design, improved performance, and rational application of UAV in spraying insecticide in rice fields.

8.1.2 Droplet Size Distribution in Rice Canopy

Grasping the regularity of droplet size distribution is of great importance to control the spray process because the droplet size is one of the most important parameters for UAV spray technology. Qin *et al.* (2016) have studied the size spectra of the droplets that displayed a typically normal distribution curve, in which fine droplets (50 to 100 μm) and coarser droplets (400 to 450 μm) account for 13 per cent and 10 per cent of the total size spectra in the upper layer, respectively. The appearance of these droplets is not desirable in the pesticide spraying process. The undersize droplet size (<50 μm) can easily drift, while the oversized droplets (>400 μm) have difficulty penetrating into the canopy layer of the crop. In the lower layer, the percentage of the droplets corresponding to the size of 50 to 100, 100 to 200, and 200 to 300 μm was 55.5 per cent, 33.7 per cent, and 8.1 per cent, respectively. The fine droplets (<50 μm) were inclined to lose their kinetic energy quickly and maintained a suspended state in air, either combined with other droplets or blown away to drift by the external wind. In contrast, coarser droplets reduced the air-borne drift loss significantly compared to fine droplet spray applications. However, coarse sprays, especially at a lower spray volume, might yield poor spray coverage, arousing growers' fears of reducing the biological efficacy of the treatment proposed a feasible way to reduce drift by changing the droplet size distribution and reducing the fraction of droplets below 100 mm. The **desired size of the droplets is in a medium range between 50 and 300 μm**, the relative content of which is approximately 59.2 per cent in the upper layer and approximately 97 per cent in the lower layer. The increase in the proportion of medium droplet size not only improves penetrability but also inhibits drift. The use of low-drift nozzles is beneficial for drift reduction. The change in droplet size generates diverse kinetic energy even if the nozzle velocity of droplets is the same. In addition, undersize and oversize droplet diameters both have difficulty landing on the target and forming a uniform disposition due to the interference of meteorological conditions and wind field by the helicopter rotor

8.1.3 Droplet Penetrability in the Rice Canopy Layer

To control the drift and to improve the adhesive rate of solution, it is crucial to enhance the droplet penetrability and obtain a homogeneous deposited distribution by optimizing the working parameters of the UAV. Qin *et al.* (2016) found that the slow flight speed (3m s^{-1}) deposited abundance of droplets in the upper layer of rice while increased flight speed (5m s^{-1}) has decreased the coverage in the upper layer of rice but increased gradually in the bottom layer. The spraying coverage in the upper layer of rice was higher when the flight height was at 0.8 m than 1.5 m. Moreover, the spraying coverage under the flight height of 0.8m reached 2.18 per cent, which was the highest. The average coverage rates in the upper layer of treatments (0.8 m, 3ms^{-1}; 0.8 m, 5m^{-1}; 1.5 m 3 ms^{-1}; 1.5m 5m s^{-1}) were 2.2 per cent, 1.6 per cent, 2.0 per cent, and 1.7 per cent, respectively. Due to the disturbance caused by the plant height and leaf area index, the coverage rate in the lower layer was smaller than that in the upper layer. The same set had average coverage rates of 0.5 per cent, 0.5 per cent, 0.6 per cent, and 0.8 per cent, respectively, in the lower layer, while the maximum value was found in treatment with 1.5 m flight height and flight speed 5m s^{-1}. This result is likely attributed to the fact that the downwash airstream generated by the UAV was conducive for the disturbance of leaves and that more droplets could reach the lower layer across the canopy. Canopy structure is one of the most important factors influencing the droplet deposition and distribution. Generally, crops in the upper and outside regions are likely to gain more depositions than are those inside the canopy. In addition, in many practical situations, the initial deposition is strongly related to spray technology, such as sprayer type, sprayer settings, and nozzle type.

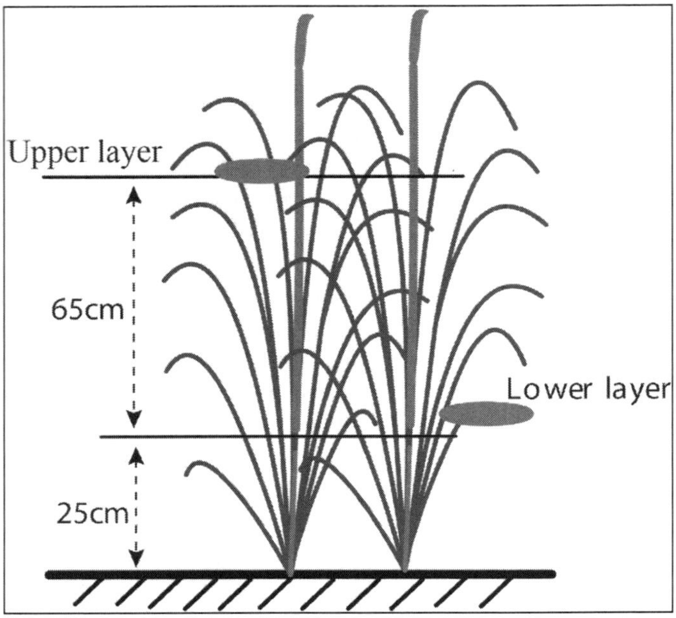

Figure 8.1: Sketches of Sampling Collectors on the Rice (*Source:* Qin *et al.,* 2016).

8.1.4 Uniformity of Droplet Distribution

The uniformity of spraying droplet distribution on targets is commonly described by the variation coefficient. The smaller a variation coefficient is, the better is the uniformity of the droplet distribution. At a spraying height of **0.8 m and a speed of 3 m s^{-1}**, the maximum value, minimum value, and average value of the spraying coverage were 6.1 per cent, 1.6 per cent, and 3.9 per cent, respectively, with a variation coefficient of 34.5 per cent in three spraying swaths. It is likely that, due to the low flight height, a combined interaction between horizontal wind and vertical wind generated by rotor wing resulted in the droplet break up and shifting. When the speed increased to **5m s^{-1}**, the maximum value, minimum value, and average value of the spraying coverage were 5.2 per cent, 2.0 per cent, and 3.8 per cent, with a variation coefficient of 29.5 per cent. The uniformity of droplet distribution improved. At a spraying height of **1.5 m and a speed of 3 m s^{-1}**, the maximum value, minimum value, and average value of spraying coverage were 5.61 per cent, 1.62 per cent, and 3.97 per cent, respectively, with a variation coefficient of 31.6 per cent in three spraying swaths. The spraying problems with overlapping and missing areas were also observed at the junction of the spraying swath. When the spraying height and speed were 1.5 m and 5 m s_1, the maximum

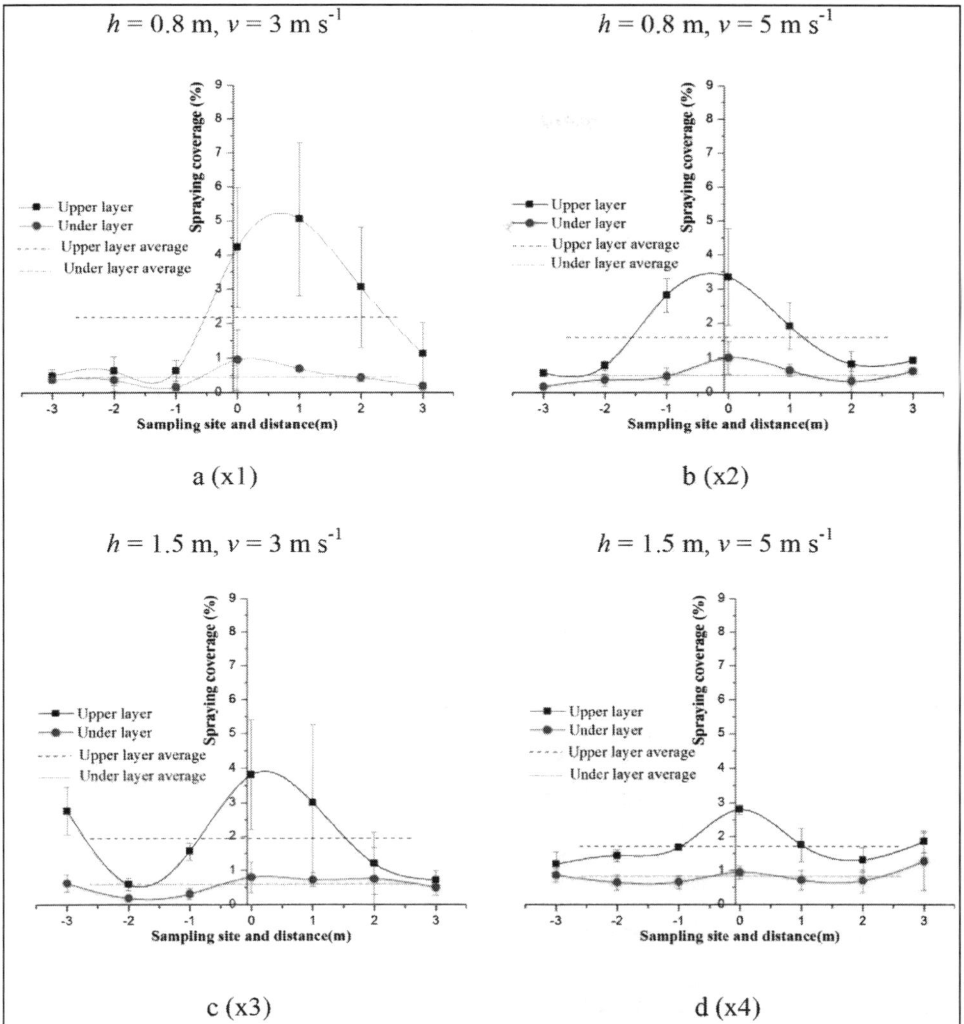

Note: During the experiment the micrometeorological conditions were: T = 27.5 ~ 29.3 °C; RH = 48.8 ~ 54.0%; Wind speed = 0.6 ~ 0.8 m s^{-1}

Figure 8.2: Distribution Diagram of Spraying Coverage under different Working Experimental Conditions (*Source:* Qin *et al.*, 2016).

value, minimum value, and average value of spraying coverage were 5.0 per cent, 2.10 per cent, and 3.7 per cent, respectively, with a variation coefficient of 23.2 per cent in three spraying swaths. Thus, when the spraying height was 1.5 m and the spraying speed was 5 m s^{-1}, the uniformity of the droplet distribution as implemented by a multi-spraying swath was optimal.

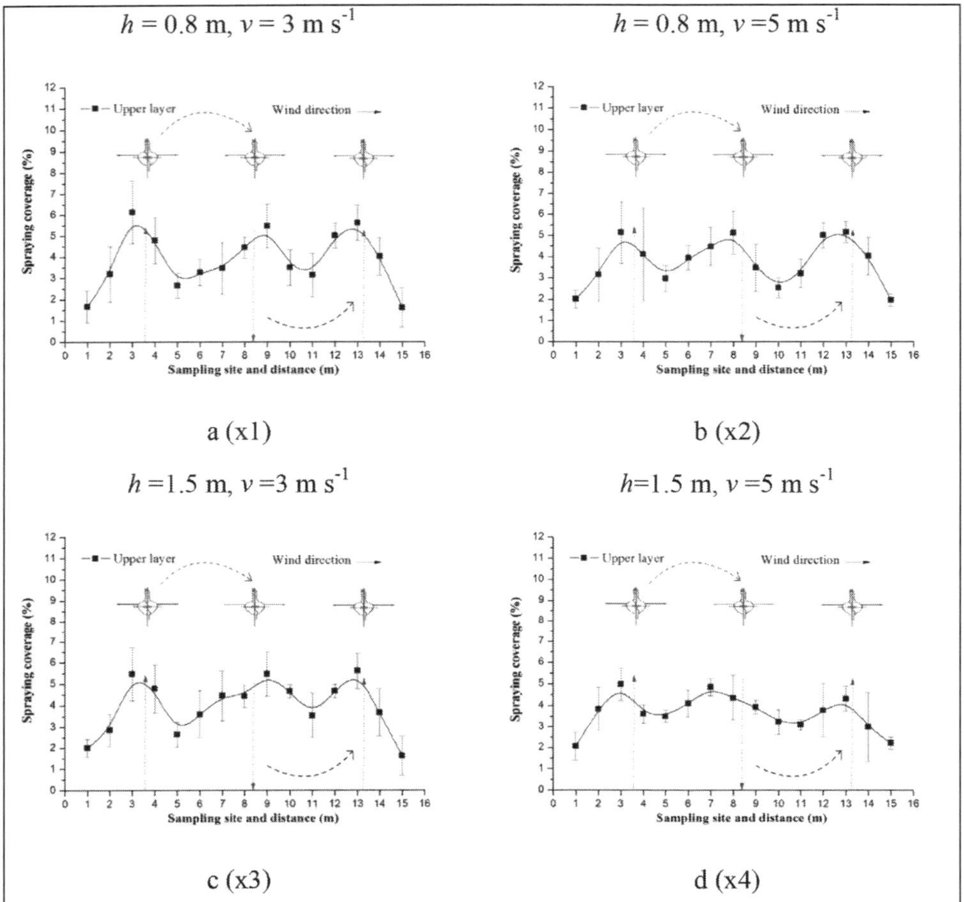

Note: During the experiment the micrometeorological conditions were: T = 27.2 ~ 28.9 °C; RH = 51.3 ~ 55.1%; wind speed = 0.4 ~ 0.7 m s^{-1}.

Figure 8.3: Droplet Distribution Curves of the Multi-Spraying Swath under different Working Experimental Conditions (*Source:* Qin *et al.*, 2016).

8.1.5 Rice Brown Plant Hoppers Control

Qin *et al.* (2016) tested the relative efficacy of UAV spray on the control of BPH in comparison to low volume and stretcher-sprayer. The control efficiency of UAV at a flight height of 1.5 m with a flight speed of 5m s^{-1} has registered highest control efficiency of 91.7 per cent on the 3rd day and continued till 10th day (74.1 per cent) while it has reduced in stretcher sprayer decreased by 23.8 per cent from the 3rd to the 10th day. In a brief summary, the controlling efficiency and duration period of the UAV greatly outbalanced those of a stretcher sprayer, where the UAV spraying of x4 showed the best control effect (91.7 per cent to 74.1 per cent) from the 3rd to the 10th day. The spraying manner of the UAV was low volume and high concentration. Moreover, the air stream that was generated by rotor wings could perturb the rice, and the droplets were prone to penetrating the canopy layer, reaching the bottom of the rice. Lee (2008) used mist and colored smoke as displaying carriers to determine the rules of the air stream of rotor wings on pesticide droplet action and successfully observed the wind field of the UAV and trajectory of pesticide droplets. However, the present description of downwash flow under the UAV on droplet disposition remains non-quantitative, especially for the vertical component of wind field, which needs to be further investigated.

Qin *et al.* (2016) have used the farm-oriented UAV (HyB-15L) to control the rice brown plant hoppers using low-altitude pesticide spraying during the heading stage of rice. They studied in detail the deposition distribution of droplets in the rice and the controlling efficiency correlated with the operating altitude and velocity. Their study has concluded the following:

- ✰ In the certain range of height (0.8 and 1.5 m) and velocity (3 and 5 m s^{-1}), the coverage rate of droplets in the bottom of the rice increased with the increased spraying height and velocity. The distribution of these droplets became more uniform when multiple spraying swaths were jointed. Meanwhile, repetitive and omitted spraying was greatly reduced.
- ✰ The spraying height played a significant role in the uniformity of deposition distribution, and the spraying velocity strongly affected the coverage ratio of pesticide droplets, especially in the bottom of the rice.
- ✰ For the control of BPH, the UAV spraying exhibited a superior efficiency than did the conventional stretcher sprayer, especially when operated at an altitude of 1.5 m and a velocity of 5 m s^{-1}. Moreover, even 5 and 10 days after pesticide application, a high controlling efficiency was still reserved, indicating that the spraying method of low volume and high concentration enhanced the duration of pesticide activity

8.2. Cotton

8.2.1 Droplet Drift

The unmanned aerial vehicles (UAVs) have unique advantage of high operational efficiency, high speed, and quick coverage. But its performance is often affected by environmental factors. When spray is undertaken with drones, a fraction of the dosage does not reach the target area which is popularly referred as droplet drift or spray drift (Kirk *et al.,* 2004). Such drift in pesticides is closely associated with wind speed and direction. Droplet deposition and drift characteristics are fundamental aspects of characterizing spray effects (Huang and Johnson, 2011).

8.2.2 Droplet Deposition in Cotton Fields

The uniformity of the spray-droplet distribution on targets is very critical in aerial sprays and it is measured as the CV (coefficient of variation) of the droplet density. The smaller the CV, the better will be the uniformity of the droplet deposition. Lou *et al.* (2018) have studied the CV of droplet distribution at different flight heights of UAV is shown in **Table 1.** At the flight height of 1.5 m, the CVs of droplet density were 117.1 per cent, 178.1 per cent, and 85.8 per cent on the upper, middle, and lower layer of the cotton canopy, respectively. At the flight height of 2 m, the CVs of droplet density were 79.4 per cent, 50.3 per cent, and 146.4 per cent. The CV of upper and middle droplet density at the flight height of 1.5 m was significantly higher than 2 m. This may be a result of the reduced flight height, which generates a strong downward swirling airflow that causes the plants to sway substantially, and the droplet density causes a significant change of the cotton canopy (Qin *et al.,* 2014).

Table 4. Distribution of Droplet Density

Working Height (m)	*Sampling Site*	*Droplet Density (Deposition/cm²)*	*Max*	*Min*	*SD*	*CV*
1.5	Upper	2.01f	6.20	0.70	2.35	117
	Middle	7.23d	30.2	0.77	12.88	178.1
	Lower	3.21e	6.40	0.73	2.76	85.8
2.0	Upper	11.35a	26.13	2.23	9.01	79.4
	Middle	9.35b	16.37	3.40	4.70	50.3
	Lower	9.17c	32.97	0.57	13.42	146.4

Note: The droplet density in the table is mean. Data followed by different small letters are significantly different among different treatments at $p < 0.05$ level by Duncan's new multiple range test.

8.2.3 Droplet Deposition Distribution Pattern

Lou *et al.* (2018) studied the droplet deposition characteristics within the spray width sprayed by UAV. The distribution of droplets in the vertical direction in the test was referred as the droplet density. At the flight height of 1.5 m, the average coverage rates were 2.5 per cent, 3.2 per cent, and 1.9 per cent for the upper, middle, and lower layers of the cotton canopy. At a flight height of 2 m, the average coverage rates were 4.9 per cent, 5.5 per cent, and 5.0 per cent. The coverage rates at 2 m were higher than those at 1.5 m. The droplet distribution in the horizontal direction in the test was represented using the deposition. At the flight height of 1.5 m, the deposition of each sampling point in the deposition area was 0.018, 0.017, 0.015, 0.027, and 0.055 μL/cm2. The deposition was least in the spray center. Downwind, deposition increases with distance from the center of the spray width. A distance from the spray width center of 2 m showed the maximum deposition. At a flight height of 2 m, the deposition of each sampling point was higher than those at 1.5 m, the maximum deposition is more than the flight height of 1.5 m. This may be due to the increase of flight height and the weakening of the downward pressure wind field below the rotor. Part droplets have a small range of drift in the downwind direction due to the lateral wind field effects. The experimental results show that the horizontal wind field and the vertical wind field below the rotor affect the distribution of droplets. Current research on wind field is not comprehensive enough and still requires further study.

8.2.4 Aerial Spray of Pesticides on Aphids and Spider Nites in Cotton

Lou *et al.* (2018) have studied the efficacy of unmanned aerial vehicles (UAVs) on cotton aphids and spider mites. They have shown that the droplet uniformity, the droplet coverage rate, the deposition, and the drifting ability are higher when the UAV flight height was 2 m. The control effects by UAV spraying on cotton aphids and spider mites were 63.7 per cent and 61.3 per cent, respectively, and the efficacy was lower than boom spraying. The data closely coincided with cotton leaf attachment and absorption of spirodiclofen after UAV spraying that were slightly lower than those after boom spraying, therefore, the control efficiency of cotton pests is slightly different. According to the different flight height operations by the UAV sprayer, the drift capability of the droplets at 2 m flight height was large, and the droplet uniformity and deposition were satisfactory. The research results provided the theoretical basis and technical support for UAV operation.

8.3. Chillies

The chilly crop is badly affected by pests and diseases causing yield reduction to the tune of 30-40 per cent. One of the most dreadful diseases is the Phytophthora blight of pepper is caused by the oomycete *Phytophthora capsici* and is one of the most destructive diseases in the world. P. capsici attacks the roots, stems, leaves, and fruits of pepper plants. The spread of the disease is very quick and rapid action is required to contain the disease at the early stage of incidence. Yet another sucking pest Aphids *Aphis gossypii* Glover is also equally devastating and reduce the yield drastically. Aphids sucks the plant sap of pepper plants causing the leaf to curl and its associated honeydew secretion attracts sooty mould infestation leading to reduced photosynthesis and yield reduction. In order to address these devastating pests and diseases in chillies, recently, Xiao *et al.* (2020) have used drones to contain them effectively, rapidly and timely management.

8.3.1 Droplet Coverage

Xiao *et al.* (2020) have shown that the droplet coverage rate of the electric air-pressure knapsack (EAP) sprayer was twice was significantly higher than that of the UAV sprayer (21.12 per cent vs. 1.83 per cent and 18.59 per cent vs. 1.43 per cent). This is because the spraying volume of the EAP sprayer (300 L/ha) was 20 times that of the UAV sprayer (15 L/ha). The amount of spray has a significant impact on the coverage of pesticide droplets, and the coverage of pesticide droplets is positively correlated with the amount of spray. The droplet coverage rate in the middle and lower parts of the second spraying was significantly lower than that of the first spraying. The main reason was that the interval between the two sprayings was 22 days, which was in the stage of the fastest growth of processing peppers. The leaf area index of chillies increased rapidly, which led to a decrease of the droplet coverage rate in the middle and lower layer of the canopy. For the double spraying, the coverage rate of the upper and middle layers was significantly better than that of the lower layer (both the UAV and EAP sprayers). Scattered growth of within the plants affects the spray fluid penetration in the plants. As many pests and diseases occur on the bottom of plants, the droplets penetrability is crucial to enhance control efficacy. The results from our study proved that the penetration of the droplets of UAV is still worse than EAP sprayer. Spray deposits decreased from the top to bottom of the canopies and decreased linearly with the increase in leaf area index.

8.3.2 Droplet Density

Xiao *et al.* (2020) showed that the spray volume had a significant impact on the droplet, and with the increase in spray volume, the droplet density showed an increasing trend. The 72 per cent cymoxanil mancozeb WP sprayed with the UAV sprayer to control *P. capsici*. The average droplet density of the upper layer sprayed with the UAV sprayer was 34.91/cm^2, that of the middle and lower layers was 23.03/cm^2 and 15.06/cm^2, respectively, and that of the ground was 9.50/cm^2. Meanwhile, the average droplet density was 60.65/cm^2, 57.16/cm^2, and 48.24/cm^2, respectively, for the upper, middle, and lowers layer sprayed with the EAP sprayer. The average droplet density of the EAP sprayer was 52.39/cm^2, which is 2.54 times that of the UAV sprayer (20.62/cm^2).

8.3.3 Uniformity of the Deposition

Xiao *et al.* (2020) showed that the uniformity of the deposition is also very important for controlling pests and diseases. To comprehensively evaluate the droplet deposition distribution in the pepper canopy, the coefficient of variation (CV) of the droplet deposition characteristics (droplet density and coverage rate) was calculated to characterize the droplet distribution uniformity after the addition of alkyl ethyl sulfonate. For low volume spraying, the relevant quality indicators could be achieved when the droplet distribution uniformity was less than 60 per cent. For UAV spraying, the CV of the droplet distribution for the upper layer of the processing pepper canopy was better than that of the middle and lower layers. Moreover, the deposition uniformity of the second spraying was significantly lower than that of the first spraying. This was due to the high canopy index of the processing pepper plant, especially in the late growth stage. The leaves were staggered and dense, which seriously affects the penetration and even the distribution of droplets. The uniformity of the deposition distribution of the UAV sprayer was influenced by many factors, such as the types, the flight accuracy, the flight parameters, the spraying system, the biased downwash wind, and the meteorological condition. Moreover, the uniformity of the droplets sprayed by the

EAP sprayer was higher than that of the UAV spraying. Wang *et al.,* reported that the deposition uniformity of the EAP sprayer depended on the stability of the traveling speed along the spraying route and on the regularity of the arm movement of the operators.

8.3.4 Droplet Penetrability

Xiao *et al.* (2020) have the studied the penetrability of pesticides when sprayed using UAV. The 72 per cent cymoxanil_mancozeb WP and 20 per cent flutamide SC was sprayed to control the *P. capsici* and the aphids. An analysis of the deposition of the upper layer of the processing pepper canopy shows that the deposition of the low-volume UAV sprayer was significantly higher than that of the EAP sprayer. The greatest deposition was achieved with the UAV sprayer with 1.01 g/cm^2, which was 98 per cent more than the EAP sprayer. The main reason for the higher upper layer of the processing pepper deposition of the UAV sprayer may be that fine droplets were better retained in the upper canopy. The droplet coverage and density of the EAP sprayer were significantly higher than those of the UAV, but the deposition rate was very low. The deposition of the three treatments sprayed by the UAV on the lower layer of peppers was 41.80–51.18 per cent that on the upper layer, which indicated that the spray had good penetration. For the second spraying, the deposition of the UAV sprayer was 2.48 times that of EPA sprayer. Remarkably, the deposition of the three treatments sprayed by the UAV on the lower layer of peppers was 16.73–25.24 per cent that of the upper layer, which was significantly lower than the first spraying. This was due to the high canopy index of the processing pepper plant, especially in the late growth stage. The leaves were staggered and dense, which seriously affects the penetration and even the distribution of droplets. The EAP sprayer showed similar results.

8.3.5 Control of the *P. capsici* and the Aphids

The control efficacy of the two sprayers applying 72 per cent cymoxanil_ mancozeb WP on peppers with *P. capsici* is indicated. The EAP sprayer showed better control efficacy on processing peppers with *P. capsici,* and the control efficacy was beyond 95 per cent 20 days after spraying.

Among the three treatments with the UAV sprayer, Treatment 2, *i.e.*, 1/2 times the field recommendation dosage, showed the worst control efficacy (81.44 per cent). The control efficacy of 1/3 times the field recommendation dosage, displayed the best control efficacy (94.34 per cent). Deposit structure plays a major role in control efficacy. According to the results of the deposition, the ESP sprayer had better deposition uniformity, which benefits the control of processing peppers with *P. capsici*. The larger area of coverage and a larger number of spray deposits increased the odds of interaction between active ingredients and pathogens. Similar to processing peppers with *P. capsici* control, the effect of the EAP sprayer on aphids was better than that of the UAV spraying. The control efficacy five days after spraying was beyond 80 per cent. It is worth noting that the control of diseases and insect pests via UAV spraying was better when the dosage of the pesticide was

1/3 times the field recommendation dosage. It was thus shown that UAV spraying has a significant effect on the reduction of pesticide application efficiency.

Xiao *et al.* (2020) have conducted a comparative study of assessing the efficiency of UAV and conventional (EAP) sprayers for pesticide applications in processing pepper fields in China. The total deposition on the plants, the uniformity of the deposition, droplet penetrability, the characteristics of the deposition, and the control efficacy of the processing pepper field with *P. capsici* and aphids were compared. They have brought out the following conclusions:

- The UAV sprayer had a poor droplet coverage rate, droplet density, and deposition uniformity. The greatest deposition was achieved with the UAV sprayer with 1.01 g/cm^2, which was 98 per cent more than that achieved with the EAP sprayer.
- Although the spray deposition characterizations of the UAV sprayer were lower than that of the EAP sprayer, the control efficacy of UAV sprayer applications of 72 per cent cymoxanil_mancozeb WP and 20 per cent flutamide SC on processing pepper fields with *P. capsici* and aphids, compared with that of EAP sprayer applications, was slightly lower.
- The growth stage of processing peppers had a significant impact on droplet characteristics and on the control e_ect of pesticides.
- UAV spraying has a significant effect on the reduction of pesticide application efficiency. The experiments demonstrated the feasibility and high efficiency of UAV sprayers in processing pepper fields. However, the deposition uniformity and the droplet penetrability of the UAV need to be improved. Due to the lower coverage and poor deposition uniformity, effective measures, such as optimizing the spraying system or adding adjuvant in the tank, which can improve deposition uniformity and penetrability, will be needed. It is also necessary to study the residue of high-concentration pesticides in pepper fruits and environments treated with a UAV sprayer.

8.4. Sugarcane

Sugarcane is widely cultivated in tropical countries and India is one of the leading producer and consumer of sugar. In the past one decade, areas under sugarcane has declined drastically particularly in Tamil Nadu, the area under sugarcane reduced from 5 lakh ha to less than 1 ha in the past 5 years. One of the prime reasons for such reasons is the requirement of labours which very scare and mechanization is order of the day from the point of planting till harvest. During the cropping period, any intercultural spray pesticides, nutrients or bioinoculants is extremely difficult due to the morphology of the crop and sharp edges of the leaf blades which often injure workers. Adding further to this some of the pests (internode borer, top borer) diseases (smut, red rot, yellow leaf disease) are highly devastating and farmers find it difficult take up any plant protection measures.

Under these circumstances, drones may be of help to deliver sprays in sugarcane fields.

8.4.1 Optimal UAV Parameters for Sugarcane

Aerial spray by drone has the advantages of high efficiency and flexibility. However, effects of spray parameters of drone on the droplet distribution on sugarcane canopy have not been reported. Zhang *et al.* (2020) conducted experiments to optimize various spray parameters (spray volume, flight height, and flight velocity) and three levels by quad-rotor drone. The **droplet deposition density** for upper layer was optimal when UAV was set at 15 L/ha of spray volume, 3 m of flight height, and 4 m/s of flight velocity. For the mid-layer and lower layers, it was found suitable using 15 L/ha of spray volume, 2 m of flight height, and 4 m/s of flight velocity. On the other hand, **droplet deposition uniformity** was optimal for the upper layer with 15 L/ha of spray volume, 4 m of flight height, and 4 m/s of flight velocity, and that 15 L/ha of spray volume, 3 m of flight height, and 4 m/s of flight velocity for the mid-layer, and 12 L/ha of spray volume, 3 m of flight height, and 5 m/s of flight velocity for the lower layer. For **droplet deposition penetration**, it was effective using 12 L/ha of spray volume, 4 m of flight height, and 4 m/s of flight velocity. Under comprehensive consideration of the density, uniformity, and penetration of droplet deposition, the optimal spraying parameters were **15 L/ha of spray volume, 3 m of flight height, and 4 m/s of flight velocity**, which could be used as a reference parameter of drone when applied in sugarcane crop.

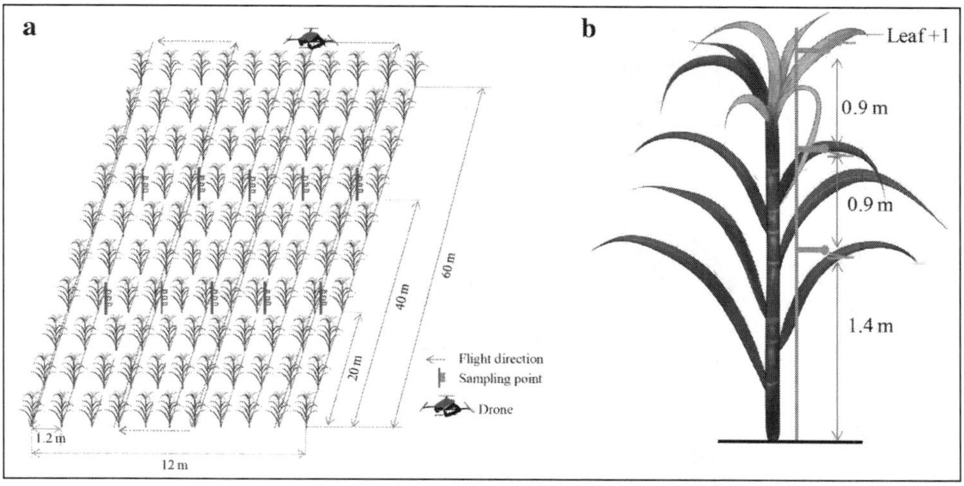

Figure 8.4: Drones Direction of Flights and Sampling Points.

8.4.2 Drones to manage Fall Army Worm in Sugarcane

The Fall Army Worm (*Spodoptera furgiperda*) is one of the invasive pests causing extensive damage in maize. The pest originally from USA and migrated to Africa in 2016 and entered in southern India in 2018. It is polyphagous pest feed on wide range cereals, millets, sugarcane, banana and other crops. It causes extensive damage in a very short span of time. In order to control the pest, quick action to be taken to ensure that pest population is kept at bay. Drones are highly useful technology to undertake spray quickly to contain the pest population.

Song *et al.* (2020) employed drones to control fall army worm in sugarcane. Drone (3WWDZ-10A, XAG) is being used to spray insecticides to control FAW to deal with the intrusion. They conducted the experiments with some marketable insecticides in Guangxi, China. The dosages were used as recommended. Water was used as control. **Flight parameters** used were velocity 3 m/s, height 3 m and spray flow 30 L/ha. The numbers of FAW before spraying and 3 days after spraying were investigatedThe population decline rate and control efficacy were calculated by the following formulas: **population decline rate** (R) = (Nb - Na)/Nb 9 100 per cent, **control efficacy** (E) = (Rt - Rc)/(1 - Rc) 9 100 per cent, where **Nb** represents the number of alive FAW before spraying, **Na** represents the number of alive FAW after spraying, **Rt** represents the population decline rate in the treatment and **Rc** represents the population decline rate in the control.

Figure 8.5: Morphological Characteristics of Fall Armyworm. (a, b) Show the 'Y' shape on the head; (c) Shows the four black flecks and d shows the thanatosis.

Figure 8.6: Damage Symptoms of Sugarcane by Fall Armyworm. (a) Shows the leaf epidermis; (b) Shows the irregular holes and incisions; (c) Shows the frass; (d) Shows the damaged whorl and ? indicates the positions ofdamage symptoms.

Song *et al.* (2020) have shown that the mixture of chlorfenapyr–chlorantraniliprole–lufenuron had the best control with 94.86 per cent population decline rate with 94.94 per cent control efficacy. The population decline rate and control efficacy of abamectin–chlorantraniliprole and chlorfenapyr–lufenuron were all over 80 per cent. However, the treatment of abamectin– dursban had only 20.82 per cent population decline rate and 21.85 per cent control efficacy. Therefore, the spray of insecticide mixture of chlorfenapyr–chlorantraniliprole–lufenuron through drone would be recommended for FAW control.

8.4.3 Use of UAVs in Spray of Plant Protection Chemicals

An extensive review was undertaken to determine the feasibility and utilization of Unmanned Aerial Vehicles (UAVs) or drones for pesticide sprays in various crops and the output from the literature is summarized in Table 2. The table clearly suggests that there are several studies have been undertaken and demonstrated successfully on the use of drones in pesticide spray in agricultural and horticultural crops. The following are the observations

1. Drone application in agriculture is primarily focused on pesticide applications. Extensive research has been done on optimization of spray volume, droplet size, spread of droplets, and penetrability besides efficacy

Table 5: Application of UAVs with the Set Parameters (Flight height, Flight speed, Nozzle Type) for Pesticides Spray in Various Crops and their Efficiencies in Comparison to Conventional Sprayers

Sl. No.	*Crops*	*Pests/ Diseases*	*UAV Type*	*Flight Height (m)*	*Flight Speed (m/s)*	*Nozzle Type*	*Spray Drift (per cent)*	*Efficiency/ Area Covered*	*Comments*	*Reference*
1	Rice	Pesticide	Four-rotor electric UAV	Maximum 1-6	Maximum 0-8	-	-	-	-	Guo *et al.*, 2019
2	Rice	Pesticide	UAV single rotor drone	1.5-3	3-6	Hollow cone, Flat cone	-	-	-	Xuan *et al.*, 2019
3	Rice	Plant Protection	Six-rotor UAV	3.5	Maximum 10	-	-	-	-	Yang *et al.*, 2017
4	Rice	-	Wind Speed Sensor Network measurement system (WWSSN) UAV	0.98	3.2	-	-	Droplet deposition in effective spray area 76.45	-	Chen *et al.*, 2017
5	Rice	Tebucona-zole	Single rotor UAV	2	5	Flow of single nozzle /L/ min -0.80	-	High efficiency of UAVs is not fully achieved	Liquid distribution was un-uniformity due to application of Pesticide in UAVs is low precision level	Wang *et al.*, 2017

Sl. No.	Crops	Pests/ Diseases	UAV Type	Flight Height (m)	Flight Speed (m/s)	Nozzle Type	Spray Drift (per cent)	Efficiency/ Area Covered	Comments	Reference
6	Rice	Brown Plant Hopper (BPH) *Nilaparvata Lugens*	Single rotor UAV	1.5	5	TEEJET nozzle	-	High control efficiency	Control of plant hoppers, the UAV spraying exhibited a superior efficiency	Qin *et al.*, 2016
7	Rice	Pesticide	Multi rotor UAV	1.5	5 with 1.9~2.5 wind speed	TEEJET nozzle	23.06	-	Droplet deposition uniformity 21.62 per cent,	Chen *et al.*, 2020
8	Rice	-	UAV	1	1	Flat fan nozzle 4nos.	5	Spray uniformity 98.5 per cent	Field performance in paddy crop was found to be 1.08 ha/h	Yallappa *et al.*, 2018
9	Rice	Fluorescent dye Rhoda-mine-B	UAV	5	3	Two fan nozzles	90 per cent drift within 8m	92.8 per cent	Use the low altitude helicopter, droplets can penetrate and liquid deposition to the lower part of crops	Xue *et al.*, 2014
10	Small bell stage Corn	Plant Protection	UAV	1	4	Two fan nozzles	-	-	-	Zheng *et al.*, 2017

Sl. No.	*Crops*	*Pests/ Diseases*	*UAV Type*	*Flight Height (m)*	*Flight Speed (m/s)*	*Nozzle Type*	*Spray Drift (per cent)*	*Efficiency/ Area Covered*	*Comments*	*Reference*
11	Corn	Plant Protection	UAV	2	2	Two fan nozzles	-	-	-	
12	Wheat	Pesticide	UAV	2	5	3 nozzle	-	effective spraying width was 5 m 16.8lit/ ha@0.4MPa	The UAV had comparable deposition and efficacy control to the EAP at a higher spray volume (>16.8 lit/ha)with coarse nozzles, but exhibited inferior deposition and efficacy control at a lower spray volume (<9.0 L ha–1) with fine nozzles	Wang *et al.*, 2019

Sl. No.	*Crops*	*Pests/ Diseases*	*UAV Type*	*Flight Height (m)*	*Flight Speed (m/s)*	*Nozzle Type*	*Spray Drift (per cent)*	*Efficiency/ Area Covered*	*Comments*	*Reference*
13	Wheat	Wheat powdery mildew	UAV	5	4	Rotary atomizer nozzle Low volume and high concen-tration	-	the droplets distribution was the most uniform, and the coefficient of variation was 33.13 per cent. control efficiency reached 55.1 per cent	the coverage rate of droplets on the wheat lower layer was the largest, as it was 45.6 per cent of that on the upper layer.	Qin *et al.,* 2018
14	Cotton	Defoliation 540 g/L thidiazuron and diuron SC: 180	UAV Spraying volume (L/ha) 15.0-22.5	2	dikhu 5	-	-	-	During 1st spray Bolls opened in field experiments in Shihezi and Hutubi are 40.74 per cent and 36.90 per cent respectively.	Yanhua *et al.,* 2019

Sl. No.	*Crops*	*Pests/ Diseases*	*UAV Type*	*Flight Height (m)*	*Flight Speed (m/s)*	*Nozzle Type*	*Spray Drift (per cent)*	*Efficiency/ Area Covered*	*Comments*	*Reference*
15	Cotton	Pesticide	ASPEE with maximum Psi up to 1400 kPa,	-	0.8333	11 m boom sprayer	-	Field efficiency 63.03 per cent,	uniformly distributed in 600 kPa Psi with swath width is 1235. The average theoretical field capacity and average effective field capacity was found to be 3.3 and 2.08 ha/h, respectively	Hunt, 1983; San-chavat *et al.*, 2018
16	Cotton	Cotton aphids and spider	UAV	1.5	-	4nos Centrifugal nozzle	5-15	cotton aphids and spider mites were 63.7 per cent and 61.3 per cent, respectively	the droplet uniformity and deposition were satisfactory	Lou *et al.*, 2018
17	Ground-nut	-	UAV	1	1	Flat fan nozzle 4nos.	5	Spray uniformity 98.5 per cent	Field performance in groundnut was found to be 1.15 ha/h	Yallappa *et al.*, 2018

Sl. No.	*Crops*	*Pests/ Diseases*	*UAV Type*	*Flight Height (m)*	*Flight Speed (m/s)*	*Nozzle Type*	*Spray Drift (per cent)*	*Efficiency/ Area Covered*	*Comments*	*Reference*
18	Soybean	Pesticide	Medium-quality sprays	-	-	Flat fan nozzle	-	Spray fluid 145lit/ha	Fine-quality sprays should be avoided for treating lower portions of a canopy unless some other form of energy, such as air-assistance spary will help move deeper into a canopy.	Derksen *et al.*, 2008
19	Sugar-cane	Pesticide	Quad-rotor drone	3	4	Positioning system and centrifugal atomizing nozzle	-	-	The efficiency applied by drone can be 60 times higher than manual spray, and the pesticide application rate is 20–30 per cent lower	Qiu *et al.*, 2020

Sl. No.	*Crops*	*Pests/ Diseases*	*UAV Type*	*Flight Height (m)*	*Flight Speed (m/s)*	*Nozzle Type*	*Spray Drift (per cent)*	*Efficiency/ Area Covered*	*Comments*	*Reference*
20	ugarcane	Fall Armyworm (*Spodoptera frugiperda*)	UAV	3	3	-	-	Control efficacy applied with chlorfenapyr–chloran-traniliprole–lufenuron were 94.86 per cent and 94.94 per cent, respectively	Drone spraying is highly efficient and flexible and has been successfully applied in sugarcane	(Lan *et al.* 2017); Song *et al.*, 2020
21	Sugar-cane	Stem Borer (Lepido-ptera)	Multi-rotor UAV	3	4	centrifugal atomizing nozzle	-	spray efficiency was 21.3 min/ hapesticide requirement was lowered as much as 90.65 per cent as compared to KES	UAV spray was found 40.0 per cent, which was higher compared to knapsack electric sprayer KES (30.0 per cent)	Zhang *et al.*, 2019

Sl. No.	*Crops*	*Pests/ Diseases*	*UAV Type*	*Flight Height (m)*	*Flight Speed (m/s)*	*Nozzle Type*	*Spray Drift (per cent)*	*Efficiency/ Area Covered*	*Comments*	*Reference*
22	Pineapple	Pesticide	UAV Oil driving single rotor Droplet size/µm-268.6 Flow rate single)/ mL/min - 800	< 2.5	1.14-2.82	-	11 per cent to < 26.44 per cent.	90 per cent spray drift distance can control in 10 m.	Effectively control the spray drift.	Wang *et al.*, 2018
23	Pineapple	Pesticide		Upto 3.5	2.02-3.59	-	55.76 per cent	90 per cent spray drift distance can be up to 33.54-46.50 m.	Total spray drift percentage also increased.	
24	Tea	Pesticide	UAV simulation platform	2	0.3	Four VP110015 flat-fan nozzles		Coverage 26.8 per cent	Increase of the UAV flight speed resulted decrease in droplet distribution uniformity.	Meiqiao *et al.*, 2019
25	Pepper	*Phytophthora capsici*	UAV	2	-	--	-	Electric Air-Pressure Knapsack EAP 21.12 per cent *vs* UAV 1.83 per cent Less	Efficiency was lesser since the spraying volume of the EAP sprayer (300 L/ha) was 20 times that of the UAV sprayer (15 L/ ha)	Qinggang *et al.*, 2020

Sl. No.	*Crops*	*Pests/ Diseases*	*UAV Type*	*Flight Height (m)*	*Flight Speed (m/s)*	*Nozzle Type*	*Spray Drift (per cent)*	*Efficiency/ Area Covered*	*Comments*	*Reference*
26	Pepper	*Aphis gossypii*	UAV	2	-	-	-	Less	”	Qinggang *et al.*, 2020
27	Grassland	5 per cent urea aqueous solution	Helicopter	2	4	hydraulic CP flat nozzles	-	62.83 per cent	-	Yao *et al.*, 2020
28	-	Pesticide	Single rotor UAV	3	5	Two flat standard nozzles	-	--		Wang *et al.*, 2016
29	-	Pesticide	UAV	-	-	Rotary Atomi-zation sprayer	-	-	Rotary atomization sprayer is more suitable for low-volume and variable-rate spraying, both are not possible by hydraulic nozzles.	Gong *et al.*, 2019

Sl. No.	*Crops*	*Pests/ Diseases*	*UAV Type*	*Flight Height (m)*	*Flight Speed (m/s)*	*Nozzle Type*	*Spray Drift (per cent)*	*Efficiency/ Area Covered*	*Comments*	*Reference*
30	-	-	Octocopter	3	1	Extended-range flat-spray nozzles	Less	30 per cent - 60 per cent	-	Joseph *et al.*, 2019
31	-	-	Octocopter	3	3	"	more	13 per cent - 22 per cent	-	
32	-	Pesticide	Six symmetrically positioned rotors	0.6	1.3	Flat fan nozzle with 0.2 MPa Psi	-	-	-	Bogus-ława and Jerzy, 2017
33	-	Pesticide	Unmanned helicopter	-	2.2	Rotary nozzles	-	-	Precision delivery of chemicals to the right place at right time	Hugan *et al.*, 2014
34	-	Plant protection	UAV	5	-	Two centrifugal rotary atomizers	-	spraying swath 5 m, their coefficient 41 per cent	The Standard for ultra-low volume spraying variation coefficient, which is less than 60 per cent	Xue *et al.*, 2016

Sl. No.	*Crops*	*Pests/ Diseases*	*UAV Type*	*Flight Height (m)*	*Flight Speed (m/s)*	*Nozzle Type*	*Spray Drift (per cent)*	*Efficiency/ Area Covered*	*Comments*	*Reference*
35	-	-	Single-rotor Unmanned Agricultural Helicopter	-	0.5	Centrifugal atomization nozzles	-	Downwash covers approximately a circle area of 3.0 m radius	-	Zhang *et al.*, 2017
36	-	Pesticide	UAV	-	2	-	-	-	-	Zhang *et al.*, 2019
37	-	Pesticide	UAV	6	2.2	Four spray nozzles, Micronair Ultra-Low-Volume (ULV) nozzles	-	30m (100ft) effective spray swath and able to spray 0.4 ha/ min (1 acre/ min)	UAV is especially good at spraying for vector control (<50-μm droplet size	Huang *et al.*, 2009
38	-	-	Fixed-wing applications	0.35m	Wind speed 0.084 (120-305 km/hr)	Flat fan nozzle	-	-	The increment of the tube pressure will strongly decrease the droplet diameter at a lower wind speed and *vice versa*	Tang *et al.*, 2016

of pesticides in insect pest control. Many of the optimization parameters were done mainly for pesticides use in agriculture.

2. Major part of research on UAV for pesticide spray in crops was carried out in rice, wheat, corn, cotton, pepper, sugarcane *etc.* as these crops consume more pesticides than any others. Further, these crops are cultivated in larger areas in contiguous blocks in developed and developing countries where drones application is feasible.
3. In order to improve insecticidal use efficiencies in crops, the drones operational parameters such as flight speed, flight height, nozzle type, payload besides drone type are be optimized for given situation. Overall, flight height of 2-3 m, flight speed of 3-5 ms-1, two fan nozzle, four rotor UAV, 15 litres payload are found optimal to undertake pesticide sprays using drones in agricultural crops.
4. Drones efficiencies are to be examined for certain application parameters such as droplet size distribution, droplet coverage, uniformity of drop lets, droplet penetrability, droplet drift and insecticidal efficiency prior to commercial use of drones in agriculture.
5. There are potential benefits of drones usage in agriculture that include large area coverage, less quantities of pesticides, labour saving, quick response time, timely operation well before pests occurrence exceeds economic threshold levels.
6. Despite the fact that there ample number advantages attached with drones technology, every country has its own regulatory guidelines for the use of drones in agriculture. Prior approvals are required from local authorities to use drones in agriculture.

Chapter 9

Drones for Plant Health Monitoring

In precision agriculture, one of the major components is the crop health monitoring includes irrigation, fertilization, pesticide sprays and timely harvest of the crop. In addition, understanding crop growth pattern is a critical component in crop monitoring (Poenaru *et al.,* 2015). In order to accomplish the task, drones are highly useful to take on site detection of problems so as to undertake corrective measures instantly. Despite it is expensive to build logarithms and establish relationship between ground truth and spectral signatures; it is user friendly after the basics studies are done. As labour availability and technical manpower are extremely constraints, drones are gaining popular in the context of smart farming. Remote sensing techniques from manned airborne and space borne systems have been widely adopted for crop monitoring purposes (Migdall *et al.,* 2009; Mulla, 2013) since measurements are non-destructive and non-invasive and enable scalable implementation in space and time (Araus and Cairns, 2014). Although many have explored crop monitoring using remote sensing technologies, they have been limited to only a small dataset or low spatial resolution images for time-series analysis. Migdall *et al.* (2009) derived crop properties such as green leaf area index, fraction of senescent material, and grain yield for precision agriculture from airborne hyperspectral imagery and satellite images. Esquerdo *et al.* (2011) studied the potential use of Normalized Difference Vegetation Index (NDVI) temporal analysis in soybean crop monitoring with Advanced Very High Resolution Radiometer (AVHRR) imagery in Brazil. Moderate Resolution Imaging Spectro-radiometer (MODIS) NDVI temporal profile was analyzed for rice to determine the health status.

In this context, several cameras are used for the plant health monitoring. Some are commonly used that include:

a. **Near infrared** to measure chlorophyll levels by showing red, green and blue depending on levels and is used for crop yield prediction.

   This can be used only in night as it interferes with sunlight. In addition, infrared is sensitive to moisture

b. **Hyper spectral camera** to see across the whole spectrum. The hyper spectral images are very large and require significant storage space. This camera is very expensive. The camera mostly suited for agricultural surveys is the Hyperspectral; it can show all the light frequencies in a single image, reducing the time and hardware needed to get all the necessary information.

The camera data can be easily analyzed using UVIRCO's in-house software. It allows them to model the data from the cameras in real-time. Depending on the type of camera used, they can see highlighted or discolored sections, which could indicate a problem. There are **specialized cameras** are used for précised measurements.

a. **NASA has UV glasses** used to monitor plant health, could be attached to drones. It cuts out chlorophyll and leaves red and blue to determine health. (Edmund optics)

b. **RAMAN spectroscopy**. A laser is used to measure plant nutrients and determine requirements. Can also detect fungi on orange trees. The laser is 1064nm.

c. **Tethered UAV** can be used to scan for longer, can be up to 70m off the ground and carry significant payloads.

d. Plants have **specific UV reflectance** that can show signs of damage or sickness. Correctly interpreting this reflectance can tell a great deal about plant health.

e. **Fractals** can be used to determine edge consistency to detect plant illness. Visible light cameras can be used for this. Point of view has a big impact on the data you will get.

In order to develop plant health monitoring drones, several strategies have been adopted.

9.1. Plant Growth Monitoring

Overall plant growth can be easily measured by plant height. There will be variations in the field as a result of fertility gradient or shading effects or due to occurrence of pests, diseases or nutrient deficiencies. **Chang *et al.* (2017)** have experimented and established a relationship between crop height and overall crop condition, irrigation, and estimation of terminal yield. In this study, a novel method was used to monitor crop height of Sorghum (*Sorghum bicolor*) using an UAV. They have acquired data seven times over the growing season and each aerial acquisition included over 200 images with significant image overlap at an altitude of 50 m above ground. Ortho-mosaic image and 3D point cloud were generated by applying the Structure from Motion (SfM) algorithm to the images. Ground control points (GCPs) were installed around the study area and they were surveyed using a real time kinematic (RTK) GPS unit for accurate geo referencing of the geospatial data products. A Digital Terrain Model (DTM) and Digital Surface Model (DSM) were generated from the 3D point cloud data, and a Crop Height Model (CHM) was then created by subtracting DTM from DSM **(Figure 9.1a and 1b).** Uniform crop grids along the center line of each variety were defined for further processing. The maximum CHM value within each individual grid was taken to represent crop height of the grid, and average of all grid heights over the whole area of each variety was calculated as crop height of individual variety. These measurements were compared with manual crop height measurements. Root Mean Square Error (RMSE) between field measurements and the proposed approach was 0.33 m. In addition, the height estimates from both field measurement and the proposed approach could be used to derive a growth curve by fitting a sigmoidal curve. The residual RMSEs between the observed and predicted value of the curve established from UAS and field measurements were calculated as 0.05 m and 0.1 m, respectively. The growth curve results showed that the proposed approach indicated less RMSE and generated more reliable growth curves for monitoring sorghum height. **Chang *et al.* (2017)** have shown that UAV data can provide non-destructive, reliable crop height measurements, and may be considered to replace the field-based crop

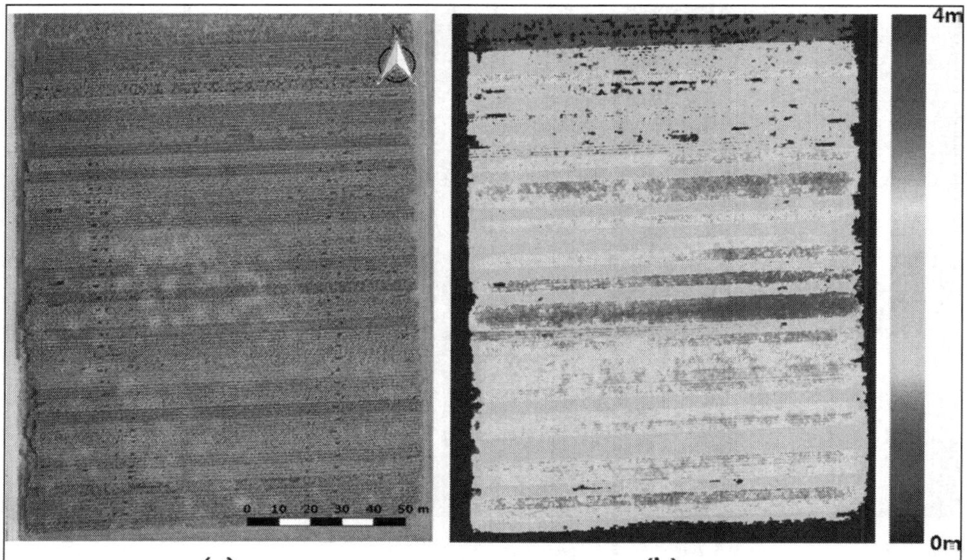

(a) (b)

Figure 9.1. Images from the Study Area of (a) Ortho-mosaic and (b) Crop height model (CHM) of SORGHUM generated from UAS data acquired on July 24, 2016 (color). (For interpretation of the references to colour in this figure legend, the reader is referred to the web version of this article).

height measurement approach. The UAV system is quite capable of crop height monitoring and could replace labour-intensive field work. In the future, UAS crop growth monitoring could be improved through temporally dense data acquisition, removal of weather effects, and other systematic errors caused from image quality and grid size.

In rice, monitoring of early stage of the crop is essential in order to determine the nutrient requirement and likely occurrence pests and diseases. This helps the farmers to save their crop at the later stages. Certainly, prevention is better than cure and farmers will be well prepared and equipped themselves to protect their crop abiotic or biotic stresses. **Norasma *et al.* (2018)** have conducted experiments to explore a possibility of using UAV in plant health monitoring at the early stage of the crop. Results showed that the RGB digital camera and multirotor UAV was successful collected the imagery at 70 m altitude and produce a tangible result in monitoring rice field in Kelantan. The imageries showed that the main problem in rice fields is the uneven ground surface and poor irrigation system that will affect the plant condition **(Figure 9.2).** The well manage and organize ground surface is a serious issue and fast action needed to make sure the paddy plot area has a good surface before planting the seed. It proved that multirotor UAV and RGB imagery can be used in rice monitoring by reducing dependence on foreign labor, minimize exposure of workers to the chemical toxic content used, access to the difficult

Figure 9.2: The Third Analysis is to Identify the Water Flow in the Paddy Plot Area. Because there's an uneven ground in the paddy plot, it will lead the inconsistent of water level condition. Since the water level should be 5 cm levelling, the plants at the area will be affected because of the flooding and their development will be stunted, consequence the declining yields production during the harvesting time. This analysis is important to make sure the water flow is in a good direction. (*Source*: Norasma *et al.,* 2018).

agriculture areas, diminish the risk of machinery being stuck or damaged during the field observation, and provide a high-resolution image in cost effective. The findings from these results, it is recommended to follow up the rice condition at every stage of the rice growth. The nutrient efficiency analysis should be conducted to check the health condition for the paddy plants in the future research by using multispectral imagery.

Recently, **Zhang *et al.* (2020)** have exploited spatial resolution of in situ UAV multispectral images to monitor the growth of rapeseed growth using a Red Edge Mica sense 3 camera installed onto a DJI M600 UAV flying at different heights to capture images of seedling rapeseed with varying ground sampling distances (GSD). The normalized difference vegetation index (NDVI) was measured by a GreenSeeker (GS-NDVI) and leaf area index (LAI) were collected to evaluate the performance of vegetation indices (VIs) and VI*plant height (PH) at different GSDs for rapeseed growth monitoring.

Zhang *et al.* (2020) indicated that NDRE had a better performance for GS-NDVI and LAI estimation than other VIs, and that the NDRE *PHDSM derived from in situ UAV images with suitable spatial resolution (1.35–2.61 cm) could achieve a higher accuracy for LAI estimation than NDRE alone. Moreover, spatial resolution is directly proportional to UAV flight height and it affects the efficiency of image acquisition and processing. The image collection and processing time at 1.35 cm GSD was about three times as long as that at 2.61 cm. Therefore, selecting a relatively low spatial resolution that ensures monitoring accuracy can reduce time and cost. The optimal spatial resolution was determined to be about 2.61 cm for rapeseed LAI estimation. The multispectral images with different GSDs were obtained through in situ collection in this study. The images with the actual GSDs were compared with images with the resampled GSDs to assess the effect of spatial resolution on rapeseed growth monitoring. This study provided an accurate and practical reference for crop growth monitoring using UAV multispectral remote sensing technology.

9.2. Plant Population Monitoring

The plant population maintenance is very critical in deciding the success of agricultural productivity. Sufficient population per unit area which is referred as the plant density is key factor deciding the productivity of corn. In order to ensure plant population at the early stage of the crop growth, UAV technology was used. In this context, **García-Martínez *et al.* (2020)** conducted experiment with five levels of nitrogen fertilization (140, 200, 260, 320 and 380 kg ha^{-1}) and images were taken at 23, 44 and 65 days after sowing (DAS) at a flight altitude of 30 m, using two drones equipped with RGB sensors of 12, 16 and 20 megapixels (Canon PowerShot S100_5.2,Sequoia_4.9, DJI FC6310_8.8). Counting was done through normalized cross-correlation (NCC) for four, eight and twelve plant samples or templates in the a* channel of the CIELAB color space because it represented the green color that allowed plant segmentation. Good precision was achieved in the first growth stages of the crop when the plants do not overlap and there are no weeds. Using sensors and unmanned aerial vehicles, it is possible to determine the emergence of seedlings in the field and more precisely evaluate planting density, having more accurate information for better management of corn fields.

García-Martínez *et al.* (2020) have clearly demonstrated that automatic counting of corn plants by comparing templates with normalized cross-correlation (NCC). This was done in the first growth stages of the crop using RGB images acquired with an unmanned aerial vehicle. Automatic counting with template comparison through NCC resulted in a counting precision of over 90 per cent up to 44 days after sowing (V5 stage), with 99 per cent precision at 44 DAS with 12 templates (samples)/120 m2. With regard to pixel size (ground sample distance, GSD) and sensors, the precision in the estimation increased as the sensor resolution increased, resulting in precision over 97 per cent and a mean error of 2.2 per cent. The results indicate that UAV technology and template comparison through NCC provide good results in corn plant counting during the early development stages of the crop, offering a good estimation of planting density in the crop. This facilitates field data compilation with high precision.

9.3. Monitoring Plant Diseases

One of the most dreadful diseases in olive plantation is a wilt caused by the soil-borne fungus *Verticillium dahliae* Kleb, is the most limiting disease in all traditional olive-growing regions worldwide. This pathogen colonizes the vascular system of plants, blocking water flow and eventually inducing water stress. **Calderon *et al.* (2013)** explored the use of high-resolution thermal imagery, chlorophyll fluorescence, structural and physiological indices (xanthophyll, chlorophyll a + b, carotenoids and blue/green/red B/G/R indices) calculated from multispectral and hyperspectral imagery as early indicators of water stress caused by wilt infection and severity. The study was conducted in two olive orchards naturally infected with *V. dahliae*. Time series of airborne thermal, multispectral and hyperspectral imagery was acquired in three consecutive years and related to wilt severity at the time of the flights. Concurrently to the airborne campaigns, field measurements conducted at leaf and tree-crown levels showed a significant increase in crown temperature (Tc) minus air temperature (Ta) and a decrease in leaf stomatal conductance (G) across wilt severity levels, identifying wilt-infected trees at early stages of the disease. Higher Tc – Ta and G values measured in the field were associated with higher VW severity levels. At leaf level, the reduction in stomatal conductance (G) caused by wilt infection was associated with a significant increase in the Photochemical Reflectance Index (PRI570) and a decrease in chlorophyll fluorescence (F). The airborne flights enabled the early detection of wilt by using canopy-level image-derived airborne Tc – Ta, Crop Water Stress Index (CWSI) calculated from the thermal imagery, blue/blue–green/blue–red ratios (B/BG/BR indices) and chlorophyll fluorescence, confirming the results obtained in the field. Airborne Tc – Ta showed rising values with a significant increase of ~2 K at low wilt severity levels, and was significantly correlated with G ($R^2 = 0.76$, $P = 0.002$) and PRI570 ($R^2 = 0.51$, $P = 0.032$). Early stages of disease development could be differentiated based on a CWSI increase as wilt developed, obtaining a strong correlation with G ($R^2 = 0.83$, P b 0.001). Likewise, the canopy-level chlorophyll fluorescence dropped at high wilt severity levels, showing a significant increase as disease progressed.

Calderon *et al.* (2013) indicated the potentials of an early detection of *V. dahliae* infection and discrimination of wilt severity levels using remote sensing. Indicators based on crown temperature such as CWSI, and visible ratios B/BG/BR as well as fluorescence were effective in detecting wilt at early stages of disease development **(Figure 9.3).** In affected trees, the structural indices PRI, chlorophyll and carotenoid indices, and the R/G ratio were good indicators to assess the damage caused by the disease.

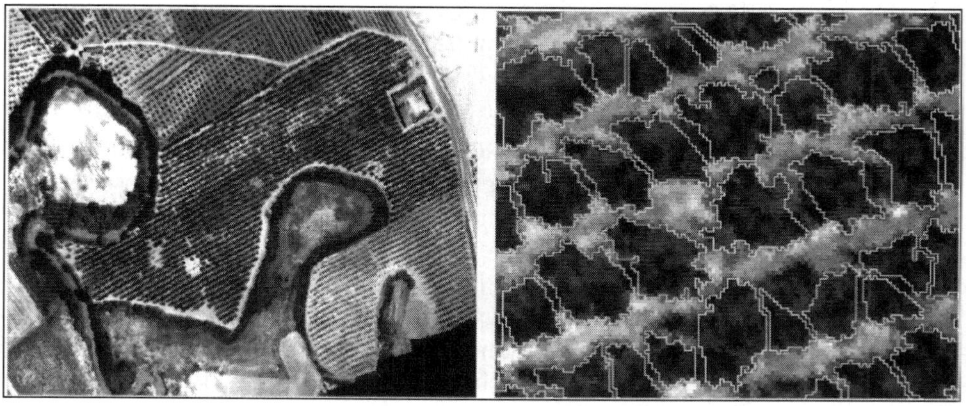

Figure 9.3: Multispectral Scene (a) obtained with the multispectral camera on board the UAV platform at 20-cm resolution, showing the Castro del Rio orchard study site (Córdoba province). (b) Automatic object-based crown detection applied to the multispectral imagery to identify pure olive crowns. Yellow square (a) shown in detail in (b). (For interpretation of the references to color in this figure legend, the reader is referred to the web version of this article.)

Nebiker *et al.* (2016) investigated the performance of new light-weight multispectral sensors for micro UAV and their application to selected tasks in agronomical research and agricultural practice. They included a high-end multispectral multiSPEC 4C camera with band pass colour filters and reference channel in zenith direction and a low-cost, consumer-grade Canon S110 NIR camera with Bayer pattern colour filters. Ground-based reference measurements were obtained using a terrestrial hyperspectral field spectrometer. Their investigations showed that measurements with the high-end system consistently match very well with ground-based field spectrometer measurements with a mean deviation of just 0.01-0.04 NDVI values. The low-cost system, while delivering better spatial resolutions, expressed significant biases. The sensors were used for crop yield estimation in rape and barley and plant disease detection in potato and onion cultivations. High levels of correlation between different vegetation indices and reference yield measurements were obtained for rape and barley. In case of barley, the NDRE index showed an average correlation of 87 per cent with reference yield. With high geometric resolutions and respective GSDs of down to 2.5 cm the effects of a thrips infestation in onion could be analyzed and potato blight was successfully detected at an early stage of infestation.

Nebiker *et al.* (2016) found significant differences between the high-end multiSPEC 4C camera and the low-cost consumer-grade S110 NIR camera. The values obtained from the high-end system consistently agree very well with the reference values with a mean difference in the order of 0.01-0.04 NDVI values. The low-cost system exhibited significant biases but showed a reasonable correlation with the values of the high-end system. The excellent results of the high-end system can be attributed to the clear separation of the spectral channels by the interference filters and the availability of a reference channel in zenith direction.

Emerging light-weight multi-head multispectral cameras for UAVs (*e.g.* Parrot Sequoia) combining a high-resolution RGB sensor with individual multispectral sensors combine the strengths of the investigated two camera types. These sensors are very promising for a number of reasons. Their high spatial resolution is valuable for geometric aspects including bundle adjustment and the extraction of geometric parameters such as crop surface models. Additionally, it allows applying UAV-based remote sensing to cost-intensive specialty crops and vegetable cultivations with thin-leafed foliage. The individual multispectral sensors with their superior interference filters and the reference channel in zenith direction, on the other hand, ensure a high quality of the spectral measurements and remove the need for ground-based spectral reference measures. Thus, UAV with suitable sensors can assist in monitoring of health status of crops which is relevant for precision agriculture. The incidence of blight infestation in potato can be resolved using NDVI mapping **(Figure 9.4).**

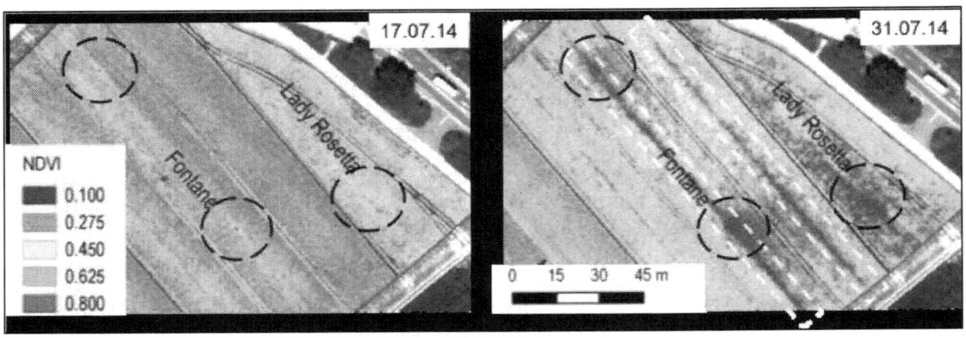

Figure 9.4. Detail View of Potato Field NDVI Map Highlighting the Three Original Sites of Potato Blight Infestation (Dotted circles) and the Subsequent Spreading of the Infestation.

The data clearly demonstrated that progression, spread and severity of the plant diseases can be assessed using multi-spectral images using drones mounted with high resolution cameras.

9.4. Monitoring Physiological Changes in Field Crops

The physiological changes in crops are to be monitored meticulously in order to optimize the correct time of fertilization, irrigation, pesticides sprays and the correct time of harvest. To address the problem of continuous monitoring of crops, UAV with sensors can be used. **Raeva *et al.* (2019)** monitored the **barley** fields using UAV and mapped several times with a multispectral and thermal sensor NDVI maps have been created for three significant months – May, June, July **(Figure 9.5).** A classification has been done with 12 classes in order to show even the smallest differences. They observed that the index for these particular months is between 0.85 and 0.97. The values for July though were the lowest. This could be due to less chlorophyll in the crop. Usually, the harvesting time of barley is April–May or September–October (Heirloom Organics, 2018). They have monitored summer barley which was greener in the early stages of growth than by the time of harvesting. This explains the lower NDVI values. The GNDVI values were similar to the NDVI ones, despite the fact that it uses the green light instead of the red one. The GNDVI is an indicator of greenness or chlorophyll content. According to this graph, the vegetation values are between 0.15 and 0.25 which is an indicator of an early harvesting time than usual. The chlorophyll content is better shown with the NDRE maps of barley **(Figure 9.6).** As stated in 4.8. the higher the NDRE values, the more chlorophyll there is. In the case of barley, the biggest NDRE values are reached in May and June. The values gradually decrease likewise the NDVI values. **Figure 9.7** is dedicated to the difference of temperature in the field. The thermal camera which was used for the study case is a long wave IR, which means that the wavelengths could infiltrate into the soil. With soil being wet, it is logical to get lower temperature values. We consider the lower temperature values with the presence of growing vegetation. After harvesting, the temperature increases.

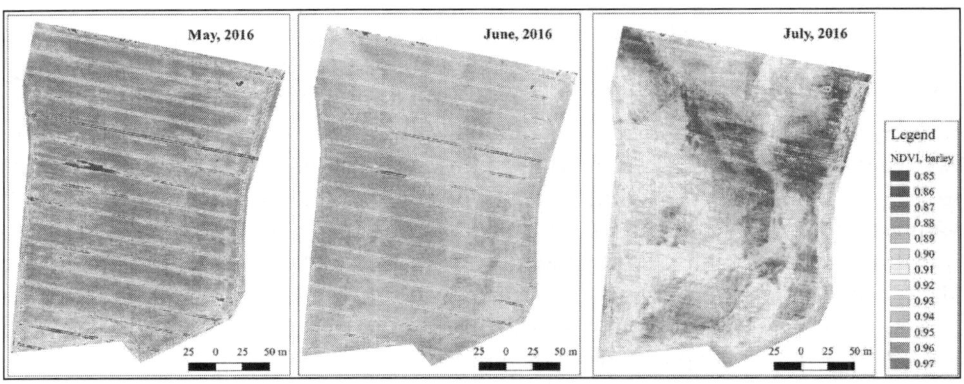

Figure 9.5. Barley: NDVI Maps, May-July, 2016.

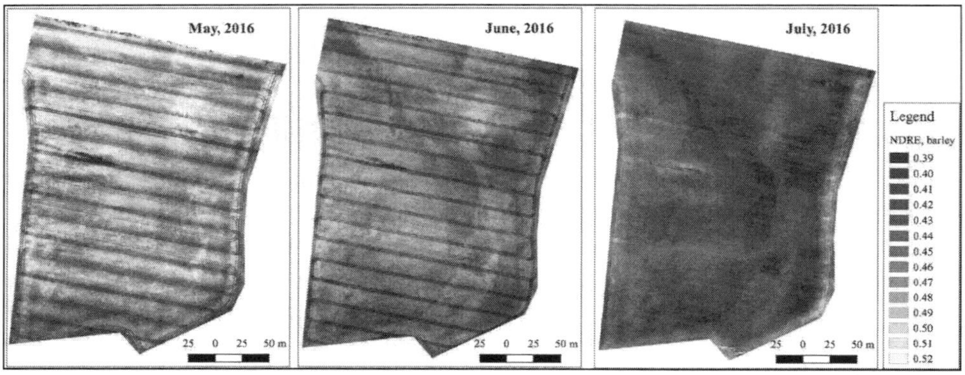

Figure 9.6. Barley: NDRE Maps, May-July, 2016.

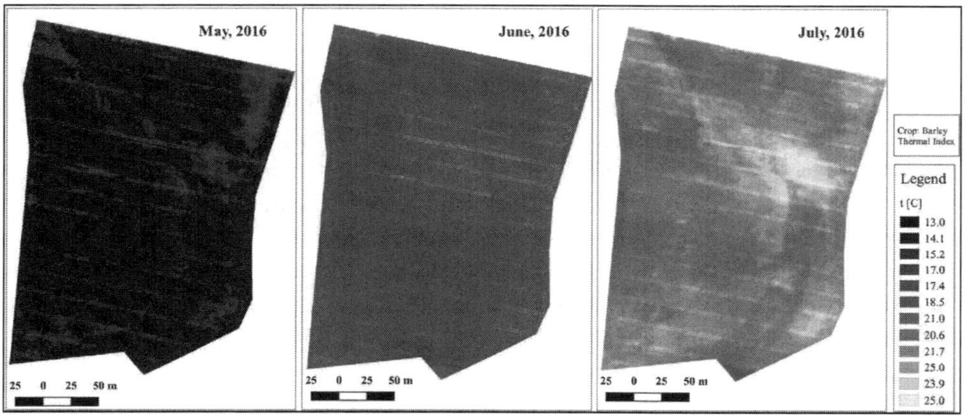

Figure 9.7. Barley: Thermal Maps, May-July, 2016.

Raeva *et al.* (2019) have used UAV for monitoring corn throughout the crop from sowing till harvest from April till May and harvested from mid-August through late October (the balance, 2018). An NDVI map has been created to show the corn development of its growing peak from May till July. The results could be seen in **Figure 9.8.** Unlike barley, corn has big green leaves and its fast-growing. The values for May and June are somewhat similar. One could notice the big increase in NDVI values from June to July. Unlike the chlorophyll presence in barley, corn is richer in chlorophyll during its latest growing stages. **Figure 9.9** depicts the increasing NDRE values where the highest are noticed in July. Again, one could notice the rapid increase between June and July. March show negative values. Having considered this graph, the authors decided not to monitor corn in the early spring. NDVI values from April to June stay surprisingly the same, whereas GNDVI is changing. July and August show really high vegetation and chlorophyll values. It is typical for corn to

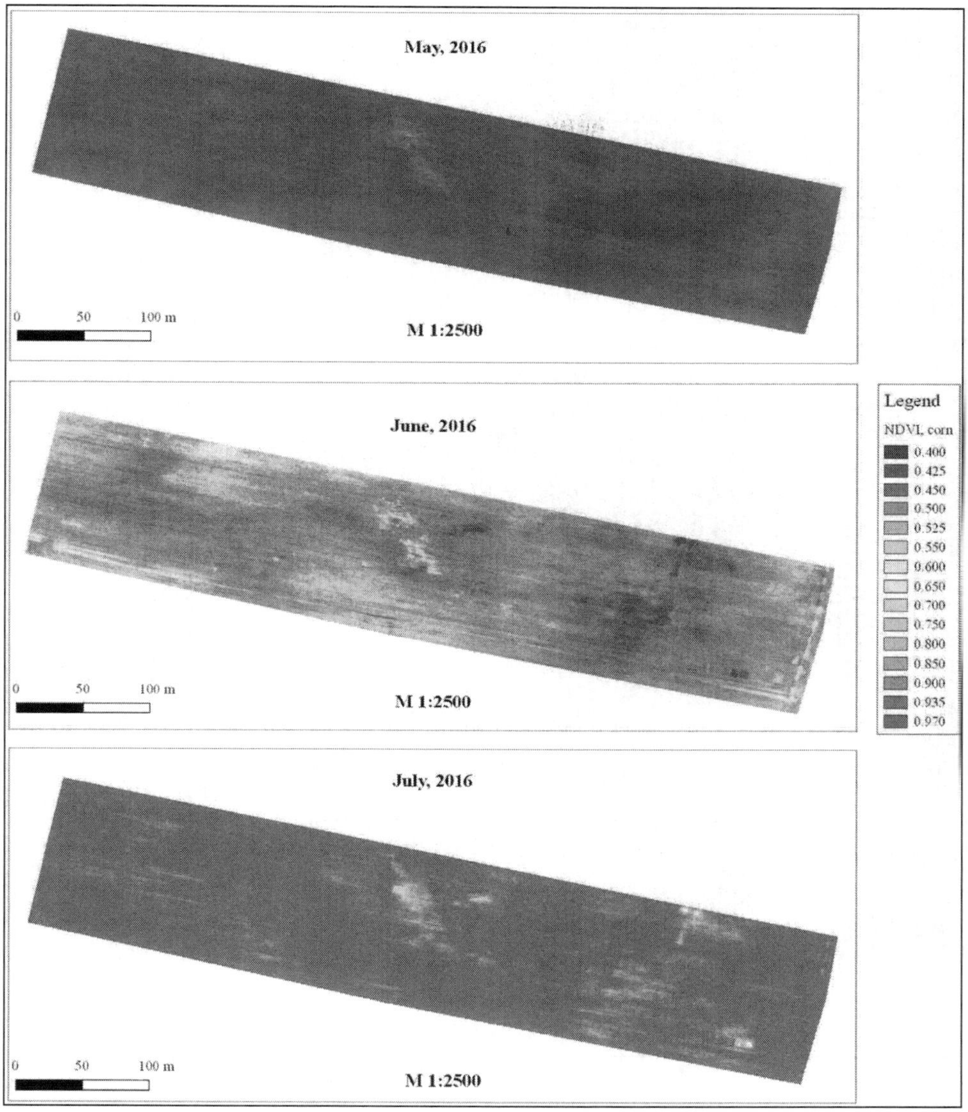

Figure 9.8. Corn: NDVI Maps, May-July, 2016.

be in its vegetation peak during that time. Due to some radiometric ambiguities, the authors decided to present the results from the corn thermal mapping only for May and June. The thermal mapping of the other months will be a subject of a new study. Nonetheless, the results could be seen in **Figure 9.10** where we notice the similar correlation between vegetation presence and temperature. In this case, the highest temperatures are in June when the NDVI and GNDVI values are higher.

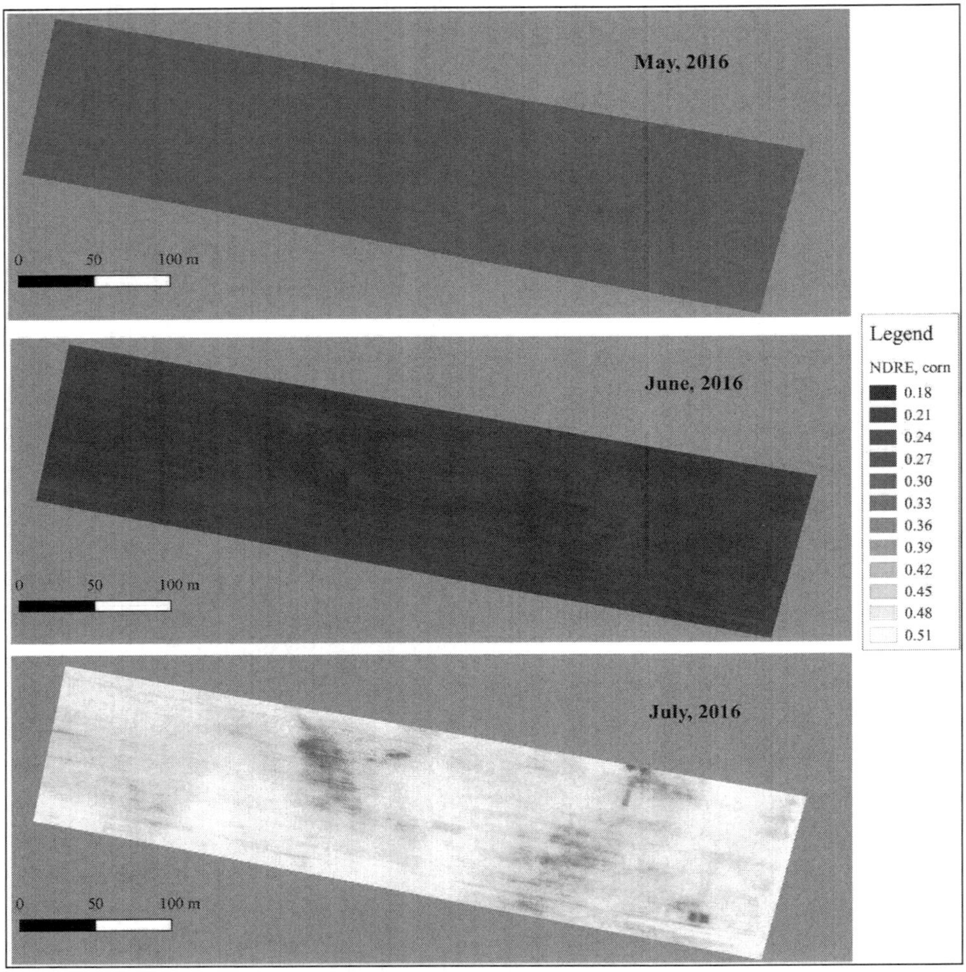

Figure 9.9. Corn: NDRE Maps, May-July, 2016.

9.5. Monitoring Physiological Stress in Forests

During the cropping period, the crops are being confronted with pests and diseases besides abiotic stress conditions. Research into remote sensing tools for monitoring physiological stress is very critical for the maintenance of the healthy and highly productive plantation forests. Forests are in generally naturally maintained with minimal or least interference by humans. In many reserve forests, it is very inaccessible to people and hard to take timely measures to overcome the physiological stresses. However, forest health monitoring using remotely sensed data from satellites and manned aircraft have become imminent. Unmanned aerial

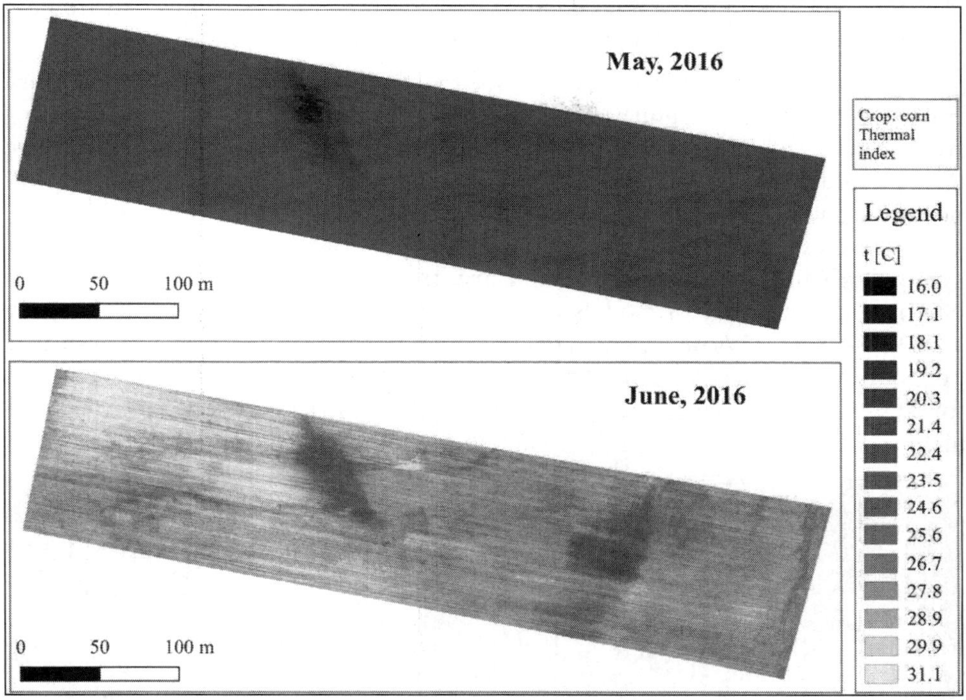

Figure 9.10. Corn: Thermal Maps, May-July, 2016.

vehicles (UAVs) may provide new tools for improved forest health monitoring by providing data with very high temporal and spatial resolutions.

In this context, **Dash *et al.* (2017)** have simulated a disease outbreak in mature *Pinus radiata* D. Don trees using targeted application of herbicide. The objective was to acquire a time-series simulated disease expression dataset to develop methods for monitoring physiological stress from a UAV platform. Time-series multi-spectral imagery was acquired using a UAV flown over a trial at regular intervals. Traditional field-based health assessments of crown health (density) and needle health (discolouration) were carried out simultaneously by experienced forest health experts.

Dash *et al.* (2017) showed that multi-spectral imagery collected from a UAV is useful for identifying physiological stress in mature plantation trees even during the early stages of tree stress. They found that physiological stress could be detected earliest in data from the red edge and near infra-red bands. In contrast to previous findings, red edge data did not offer earlier detection of physiological stress than the near infra-red data. A non-parametric approach was used to model physiological stress based on spectral indices and was found to provide good classification accuracy (weighted kappa = 0.694). This model can be used to map physiological stress based on high-resolution multi-spectral data.

Results from this study clearly showed that collection of multispectral imagery from a UAV provides a useful method to detect and monitor the symptoms resulting from a simulated forest disease outbreak resulting in foliar discolouration **(Figure 9.11).** The nonparametric models created from this data predicted the onset of needle discolouration well and had a high overall accuracy (kappa range 0.39–0.64). Analyses showed the optimal spatial resolution of imagery to range from 1 m for symptoms expressed on both small and large tree clusters. Further research should compare the accuracy of models created from high resolution satellite imagery with those from UAVs in detection of physiological stress resulting from disease or pest outbreaks. They concluded that UAVs may have an important role to play in forest health monitoring, filling the spatial and temporal gap between large-scale airborne assessment and infrequent small-scale ground surveys.

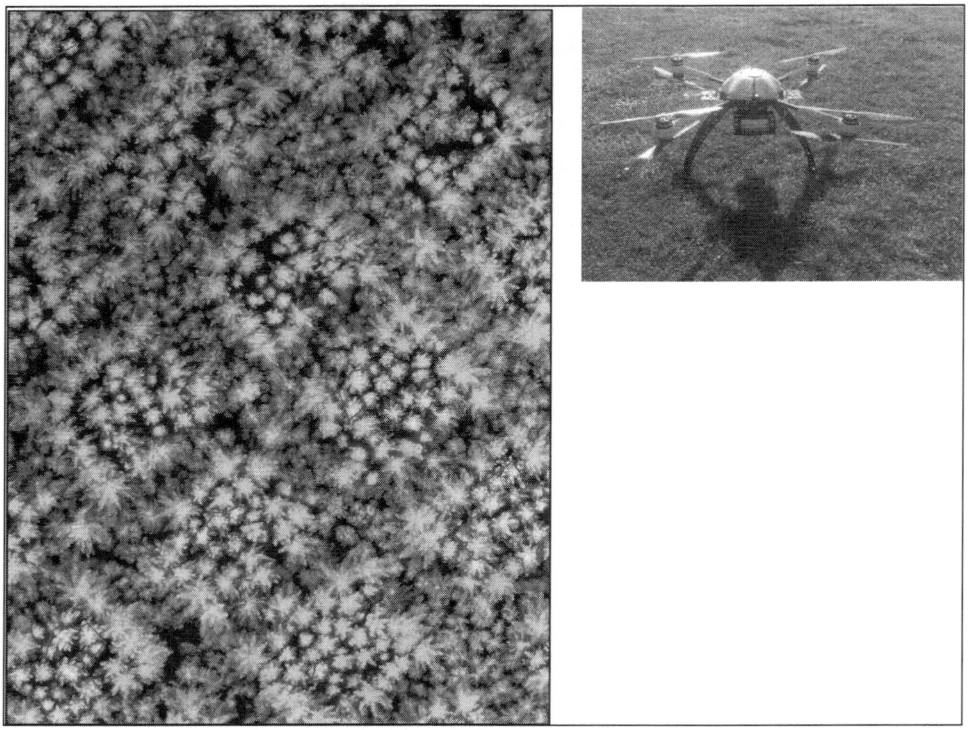

Figure 9.11. Location of the Study site within New Zealand and an Image of the Coaxial Quad-copter UAV used during Data Collection. The main panel shows an image of the multispectral data collected with treated trees evident and outlined in blue (image date = 03-03-2016). (For interpretation of the references to colour in this figure legend, the reader is referred to the web version of this article) (*Source*: Dash *et al.*, 2017).

The success of forest surveillance is largely governed by being able to match the temporal, spatial and spectral characteristics of symptoms with an appropriate platform, sensor and image analysis method. Over the past two decades high resolution satellite imagery has not had an adequate temporal resolution for routine monitoring. As a result, airborne platforms have been more widely used as they are more flexible and responsive (McConnell *et al.,* 2000). The advent of low-cost, easy to use UAVs and miniaturized multi-spectral sensors may change the operational strategy for forest health monitoring. Our results suggest a role for UAVs as a means of filling the gap between the large-scale capabilities of airborne surveillance and the need for targeted, regular surveillance of smaller areas in a cost-effective manner. In contrast to orbital satellite platforms, UAVs can also supply imagery on cloudy days which can be an important consideration in areas where overcast conditions are common during periods of peak symptom expression. Compared to both satellite and manned aircraft platforms, the time and financial resources needed to prepare and initiate a UAV flight is greatly reduced, allowing for greater flexibility in scheduling imagery acquisition.

9.6. Monitoring Forestry Nurseries

Nursery production on a large scale as in the case of forest nurseries requires using the latest solutions in the field of yield diagnostics and evaluation of the health condition of plant material. The development of modern mapping techniques allows the use of remotely controlled devices to implement unconventional analyses and presentations, *e.g.* photographic scan of the vegetation useful to carry out planning and spatial visualizations, enabling the presentation of any object visible on the ground surface. In addition, 3D scan effect more and more often used in the post-processing ensures accurate representation of a real object or object with complete planimetric and stereometric characteristics (Gamon *et al.,* 1995, Carlson *et al.,* 1997, Sullivan *et al.,* 2004, Grenzdörffer *et al.,* 2008, Dunford *et al.,* 2009, Mäkynen *et al.,* 2012, Smith 2015).

Another attribute is the ability to develop a final manipulation of the colour and rendering, especially useful in 3D graphics with studies that require filtering atmospheric phenomena occurring in the ground-level air layer, *e.g.* radiation fog, mist, *etc.* Possibilities to use precise photography in diagnosing phenomena taking place within the vegetation seem to be endless. Nowadays, nurserymen use methods of analyzing high resolution photos to monitor defoliation caused by biotic stress factors, degree of damage to assimilative apparatus due to hail or other abiotic factors, including those of anthropogenic origin. The same techniques are used in the assessment of plants' demand for water, foliar fertilization effects and finally to analyse the results of protective treatments. Large area monitoring of the land leaf area in the nursery farm and the evaluation of the percentage of high plants covering the ground are increasingly used. In addition, one can take advantage of a number of techniques for assessing the health condition of plants using the ability of assimilative apparatus to absorb and reflect solar radiation such as NDVI.

(Normalized Difference Vegetation Index), SAVI (Soil Adjusted Vegetation Index), ENDVI (Enhanced Normalized Difference Vegeta-tion Index), OSAVI (Optimized Soil Adjusted Vegetation Index) also RDVI (Renormalized Difference Vegetation Index) (Ceccato 2001, Tran and Grishko 2007, Kim *et al.*, 2015).

Durlo *et al.* (2015) have analyzed the spatial data and physiological characteristics of the vegetation cover in a nursery producing forest nursery stock with the use of the container technology. They conducted a research in the production fields of "Nêdza" nursery farm in Rudy Raciborskie Forest District located in Silesia, Southern Poland. They took advantage of DJI Phantom 3 Advanced quadrocopter equipped with a high resolution video camera, flight control system and positioning systems to perform the research. The analysis of photographic material was made on the DroneDeploy platform as a part of Map Engine Processing. Based on high resolution images, the analysis of the production area structure was performed, type of forest nursery stock was inventoried, height parameters of seedlings and degree to which the soil is covered with leaves were specified as well as normalized difference vegetation index (NDVI, RGB/VARI) was calculated. The data suggest that UAV technology attached with a video camera assist in the determination of health status of forestry nursery.

Drones mounted with sensors can assist in monitoring of plant population, early stage of establishment, crop growth, nutrient status, weed menace, incidence of pests and diseases and other vital information. This will serve as a data base to implement digital farming. The published literature clearly demonstrates that drones application in plant health monitoring is very important and it will reduce the response time to take remedial measures to protect the crop from pests and diseases.

9.7. Monitoring plant growth through various indices

9.7.1. Normalized Difference Vegetation Index

The normalized difference vegetation index (NDVI) is a simple graphical indicator that can be used to analyze remote sensing measurements, typically, but not necessarily, from a space platform, and assess whether the target being observed contains live green vegetation or not. The NDVI value ranges between - 0.70 and +0.69. The NDVI value above zero of the wetland farm was reclassified at an equal interval of 0.1 for easy interpretation. An NDVI close to 0 corresponds to no vegetation, while NDVI close to +1 (0.8 - 0.9) (Table 2) indicates the highest possible density of green leaves. The NDVI value from 0 to 1 indicates the relative greenness of the vegetation (see Fig.9.12). The relatively lesser greenness in the image indicates the harvest stage of paddy crop in the fields. The NDVI class of 0.5 to 0.7 indicates the relatively higher greenness of the leaves and the very good health of the crop.

The Minimum and Maximum value of TNAU Wetland farm generated by different indices

S.No.	*Vegetation Index*	*Minimum*	*Maximum*
1.	NDVI (Normalized Difference Vegetation Index)	-0.70	+0.70
2.	Normalized Difference Vegetation Index Red Edge	-0.50	+0.30
3.	Blue-Normalized Difference Vegetation Index	-0.70	+0.60
4.	Normalized Green–Red Difference Index	-0.65	+0.60
5.	Green Difference Vegetation Index (GNDVI)	-0.40	+0.40

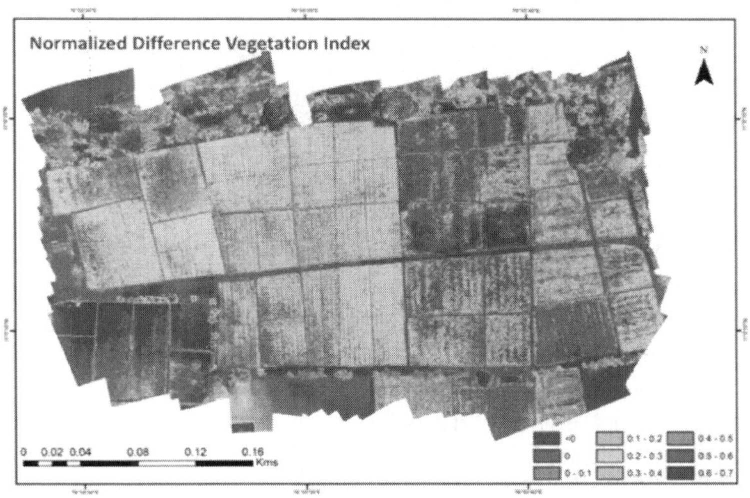

Fig.9.12. NDVI map of Wetland farm, TNAU, Coimbatore

9.7.2. Normalized Difference Vegetation Index Red Edge (NDRE)

NDRE index was formulated using the Red edge band is available in a sensor. It is sensitive to chlorophyll content in leaves (how green a leaf appears), variability in leaf area, and soil background effects. From the image it could be inferred that the NDRE value ranges from -0.5 to 0.3(see Table)(see Fig.9.13). The NDRE value between 0.2 to 0.3 indicates the presence of higher chlorophyll content in the leaves and lesser influence of soil characteristics and also from the image it could be inferred that the denser tree canopy is coming under this class. The value lesser that zero clearly shows the absence of chlorophyll and the effect would be due the barren land or dry soil exists in the farm. However, the higher greenness exhibited NDVI value for Daincha fields shows lesser NDRE values and it ranges between 0 and 0.1.

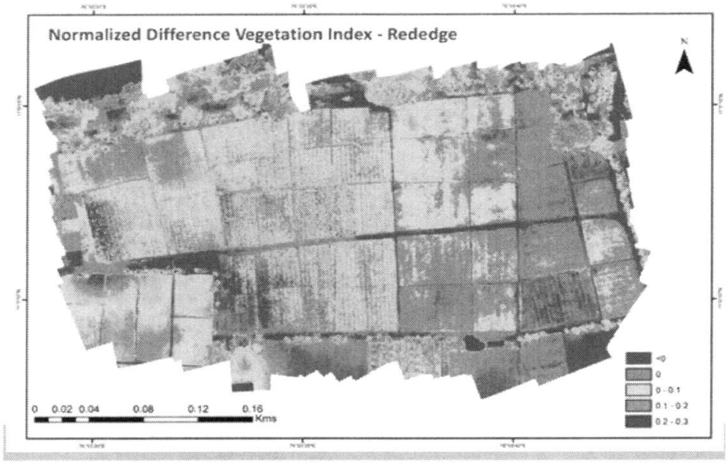

Fig.9.13. NDVI-Rededge map of Wetland farm, TNAU, Coimbatore

9.7.3. Blue-Normalized Difference Vegetation Index (BNDVI)

The Blue-Normalized Difference Vegetation Index was generated using the NIR and Blue wavelength region and the value ranges from -0.7 to 0.6 (see Table) (see Fig.9.14). The index shows the absorption of blue light by anthocyanin and scattering of blue light at the surface of the crop canopy and thereby we could infer the level of crop stress. However, the BNDVI values shows lesser contrast between stressed versus non-stressed crops. Similar to other indices, the value below zero indicates the paddy crop at harvest stage and the barren soil.

From the data it could be confirmed that the BNDVI is less sensitive to differentiate stressed crops when compared to NDVI and it also failed to differentiate the crop with soil or barren fields. From the study, it could be concluded that BNDVI may be used as an alternative in the absence of multispectral sensor.

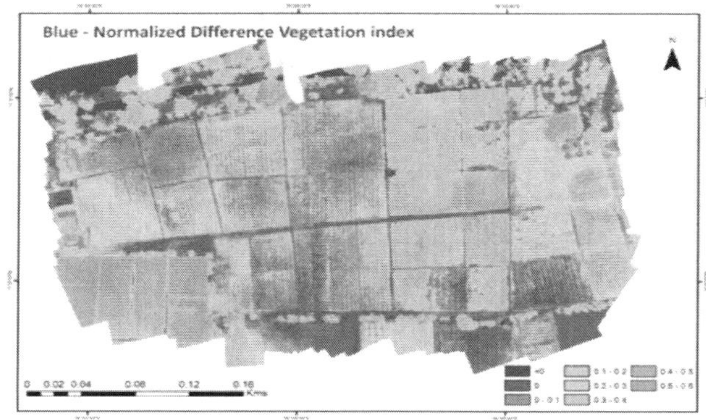

Fig.9.14. Blue-NDVI map of Wetland farm, TNAU, Coimbatore

9.7.4. Normalized Green-Red Difference Index (NGRDI)

The main advantage of NGRDI is that the index uses visual spectrum of wavelength range for the calculation and hence it is easily interpreted by our human eye. The index value ranges between -0.65 to 0.6 (Table). NGRDI value of zero becomes could be considered as an effectivethreshold to distinguish between green vegetation and other types of the ground covers (see Fig.9.15). We could infer from the images; single date image is less useful to discriminate between the objects.

The NGRDI was meant to monitor the phenology of the crop, from the image it could be clearly captured the different stages of Daincha in the field.

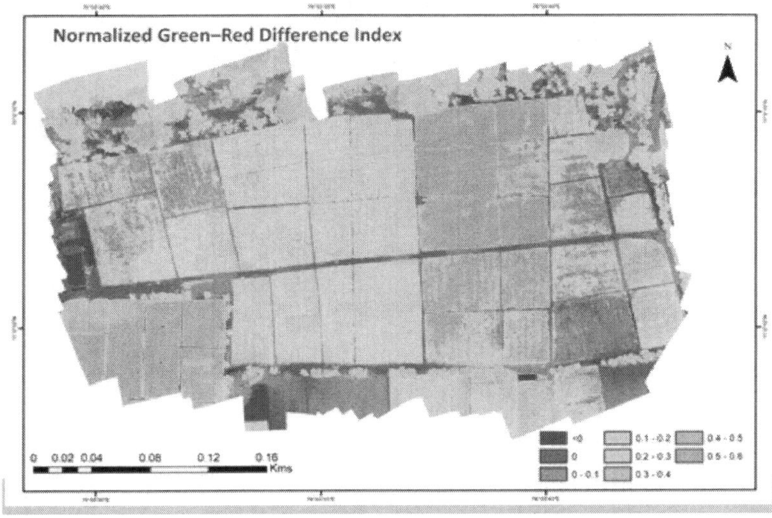

Fig.9.15. Normalized Green-Red Difference Index map of Wetland farm, TNAU, Coimbatore

9.7.5. Green Difference Vegetation Index (GNDVI)

The Green Normalized Difference Vegetation Index (GNDVI) is an index of plant "greenness" or photosynthetic activity. It is one of the most commonly used vegetation indices to determine water and nitrogen uptake into the crop canopy. The GNDVI values ranges between -0.4 and 0.4. The GNDVIvalue lesser than zero indicates the absence of canopy moisture or soil moisture in the fields. The values above zero indicates the relative increase in crop canopy moisture (see Fig.9.16).

From the image, it could be inferred that the transplanted paddy fields and the trees with denser vegetation are having the values between 0 and 0.4 and indicates the presence of canopy moisture in the fields. However, the index failed to discriminate the soil moisture and crop canopy moisture. The index fails to discriminate within field variation in the crop canopy moisture of rice fallow Daincha fields.

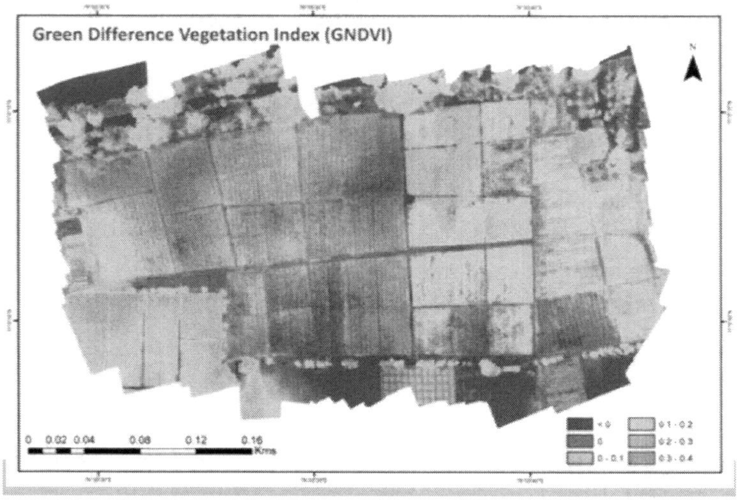

Fig.9.16. Green-NDVI map of Wetland farm, TNAU, Coimbatore

Chapter 10

Drones for Harvesting and Yield Assessment

Remote sensing (RS) is one of the fast emerging technologies which can obtain the information about an object without making physical contact with the object and it can efficiently obtain canopy spectra data in a non-destructive way, which carries valuable information indicating the canopy interaction with solar radiation such as vegetation absorption and scattering (Thenkabail *et al.*, 2011). The vegetation canopy spectra are closely related to vegetation growth. In the **visible range**, vegetation has strongly absorption of light due to the pigments and presents as low reflectance (Woolley, 1971). But in the **near infrared** (NIR) range, vegetation reflectance is relatively high and affected by thick plant tissues and canopy structure (Gausman *et al.*, 1969).

There is a close correlation established between vegetation spectra and growth parameters of plants such as chlorophyll content, leaf area index (LAI) and biomass and thus a lot of vegetation indices (VIs) calculated from reflectance of different spectra ranges (Hatfield *et al.*, 2008) have been proposed to accurately estimate these parameters. For example, Peng *et al.* (2013) used Enhanced Vegetation Index (EVI) and Wide Dynamic Range Vegetation Index (WDRVI) obtained from MODIS to accurately estimate the gross primary productivity in crops with coefficients of variation below 20 per cent in maize and 25 per cent in soybean (Peng *et al.*, 2013). Li *et al.* (2014) applied several red-edge based spectral indices to estimate plant nitrogen uptake with the coefficient of determination above 0.76. Parametric statistical approaches based on VIs are by far the simplest and most studied variable estimation approaches and have been widely used in monitoring crop growth (Verrelst *et al.*, 2015).

The changes of crop growth status monitored by spectral measures can be directly used to determine its ultimate yield. Therefore, VIs have also exhibited good potential in remote estimation of crop yield especially at the large scales (Rahman *et al.*, 2012; Sun *et al.*, 2017). Becker-Reshef *et al.* (2010) estimated the wheat yield in Kansas and Ukraine with a 7 per cent error by using time series Normalized Difference Vegetation Index (NDVI) data from MODIS. Holzman *et al.* (2014) came up with a new method to estimate regional crop yield using the Temperature Vegetation Dryness Index (TVDI) with the RMSE values ranged from 12 to 13 per cent for soybean and 14 to 22 per cent for wheat. Sakamoto *et al.* (2014) mapped U.S. corn yields successfully using Wide Dynamic Range Vegetation Index (WDRVI) derived from time series MODIS data with the estimation error below 30 per cent at the state level. Generally, VI-based methods are the mainstream approach for crop yield prediction, and a lot of regression algorithms of using VIs to estimate crop yield were established including simple linear functions and complex non-linear functions (Tucker *et al.*, 1980). Many experiments showed that the use of appropriate VI was the key to crop yield estimation instead of the complex function structures in VI-based methods (Gholizadeh *et al.*, 2015). Therefore, it is particularly important for crop yield evaluation to get the exact VI values.

Remote sensing technologies are widely used for predicting the time of harvest and quantification of crop yields well before harvesting is done. This will help the farmers to take decision on harvest and market the produce. Since the price for agricultural produce is highly volatile and fluctuating, the remote sensing technologies become very appropriate and relevant to the current production scenarios. In order to assess the crop yields, measurements such as canopy reflectance, infrared canopy temperature, and LiDAR- or SONAR-based estimates of height. Despite the technologies is cost prohibitive at the beginning, now it becomes affordable (Mulla, 2013). Spectral vegetative indices (SVIs) are functions of canopy reflectance developed to assess ground cover, crop health, drought stress, and nutrient deficiencies in several major crops. These remote sensing tools can be integrated into UAV to enable real-time assessments of crop nutritive value parameters at the field scale (Zhang and Kovacs, 2012). The canopy reflectance to predict nutritive value in alfalfa monocultures (Starks *et al.*, 2016) and perennial forage grasses (Starks *et al.*, 2006). Starks *et al.* (2016) report predictions based on canopy reflectance to explain 77–83 per cent of variation in crude proteins, neutral detergent fibre (NDF), and animal digestible fibre. Further assessment using ultrasonic or LiDAR (Light Detection and Ranging) technology enable remote measurement of crop height and facilitate improved estimations of crop biomass (Pittman *et al.*, 2015).

10.1. Rice

Duan *et al.* (2019) explored the use of remote sensing to acquire high spatio-temporal resolution imagery on a regional scale in China. Earlier reports of the same team have shown that the predictive ability of vegetation index (VI) decreased under the influence of panicle emergence during the later stages of rice growth. **Duan *et al.* (2019)** exploited a new approach which integrated UAV-based VI and abundance information obtained from spectral mixture analysis (SMA) which was established to improve the estimation accuracy of rice yield at heading stage. The six-band image of all studied rice plots was collected by a camera system mounted on an UAV at booting stage and heading stage respectively. And the corresponding ground measured data was also acquired at the same time. The relationship of several widely-used VIs and Rice Yield was tested at these two stages and a relatively weaker correlation between VI and yield was found at heading stage. The reason was the uneven emergence of rice panicle at heading stage which caused the decrease of predictive ability of VI for rice yield. In order to improve the accuracy of yield estimation at heading stage, a fully constrained least squares linear spectral mixture method was used to eliminate the influence of the panicle appearance on yield estimation. The abundance images of six mainly end members in paddy field was produced based on the six-band UAV image and ground measured spectra, including top layer leaf, bottom layer leaf, top layer panicle, bottom layer panicle, dry soil, and wet soil. In order to improve the estimation accuracy of rice yield at heading stage, the plot-level abundance of panicle, leaf and soil, indicating the fraction of different components within the plot, was derived from SMA on the six-band image and in situ end member spectra collected for different components. The results showed that VI incorporated with abundance information exhibited a better predictive ability for yield than VI alone. Among the test VIs, NDRE, and GNDVI multiplied by the difference of leaf and panicle abundance were the most accurate for yield estimation in rice under different nitrogen fertilizer treatment with least estimation error. And the product of VI and the difference of leaf abundance and panicle abundance was the most accurate index to reliably estimate yield for rice under different nitrogen treatments at heading stage with the coefficient of determination reaching 0.6 and estimation error below 10 per cent.

10.2. Maize

Wahab *et al.* (2018) have used remote sensing methods to assess maize (*Zea mays* L.) vigor and yields has had limited applications in Sub-Saharan Africa (SSA) due largely to limitations associated with satellite images. The study employed a quadcopter flown at 100 m over farm plots and equipped with two consumer-grade cameras, one of which is modified to capture images in the near infrared. We find that UAV-derived GNDVI is a better indicator of crop vigor and a better estimator of yields–r = 0.372 and r = 0.393 for mean and maximum GNDVI respectively at about five weeks after planting compared to in-field methods like SPAD readings at the same stage (r = 0.259).

The Normalized Difference Vegetation Index (NDVI) is the most commonly used vegetation index, first postulated by Rouse *et al.* (1973), quantitatively measures vegetation conditions. It is, essentially, a ratio of the difference and sum of the reflectance in NIR and red bands. However, rather than use the NIR and red bands which are the traditional band combination for NDVI extraction, we use the NIR and green to calculate green normalized difference vegetation index (GNDVI). At lower altitudes, there is no significant scattering of shorter wavelength light and so green performs equally well.

Thus,

$$GNDVI = (NIR - G)/(NIR + G)$$

where,

NIR: Reflectance values in the near-infrared band

G: Reflectance values from the green band.

The variances in the reflectance properties of the NIR and green bands

enable us to assess density and intensity of vegetation using the reflectivity of solar radiation. The utility of a VI such as the NDVI rests on its high correlation with biophysical parameters of plants and low sensitivity to others such that signals from vegetation is enhanced while minimizing solar irradiance and soil background effects (Jackson *et al.,* 1991). Normalization is achieved by denominating the difference of the NIR and G bands by their sum and by so doing becomes comparable over time and across seasons. Bausch *et al.* (2008) argue that reflectance at green wavelengths is more responsive to variations in leaf chlorophyll and plant health. In concurrence, Burke and Lobell (2017) point out that using the green band is more likely to capture differences in nutrient deficiency that are correlated to yields. Thus, plant vigor is captured by the amount of fluorescence of NIR light in a relatively cheaper sensor with a band order of NIR, green and blue. Thus earlier equation for NDVI has been transformed into

$$GNDVI = (Band\ 1 - Band\ 2)/(Band\ 1 + Band\ 2), \quad (2)$$

Wahab *et al.* (2018) demonstrated that GNDVI derived from UAV imagery is a reliable and timely predictor of crop vigor and yields and that this is applicable even in complex smallholder farms in Sub Saharan Africa.

The GNDVI derived from UAV is not only useful in ascertaining maize crop vigor and but also estimating yields and that it indeed performs comparatively better when compared to in-field methods of SPAD meter measurement and visual scoring of crop vigor. Thus, UAV-derived GNDVI is a better, most timeous, more efficient, and objective measure of crop vigor and predictor of maize yields compared to in-field methods. The GNDVI method as used here is also the most desired if the objective is to provide decision support for smallholder farmers given that it gives these results at the earliest–5 WAP (weeks after planting); compared to SPAD measurements–9 WAP; and visual scoring–13 WAP. It thus could afford farmers the opportunity to take remedial actions to improve final yields by targeting critical farm management activities such as fertilizer application and irrigation rather than treat whole plots as homogeneous units. The relatively poor performance at 5 WAP and even 9 WAP of the visual scoring approach to assessing crop health and thus expectations with regards to yields by close of the farming season implies that it is not an approach that can be relied on by farmers to take farm decisions. Its relatively improved performance close to the season's end eliminates its utility as not much could be done to remedy a poor season. The relatively better performance of the UAV-derived GNDVI as early as 5 WAP is the highlight of this study and bodes well for improving yields even on complex smallholder farms in SSA.

10.3. Maize

Geipel *et al.* (2014) used spectral and spatial information from UAS-based RGB imagery for predicting corn grain yield in early- to mid-season crop growth stages. RGB imagery was used to compute crop surface models and to extract crop height information. They tested three linear regression models for the prediction of corn grain yield with R^2 determination coefficients of up to 0.74 and RMSEP ranging from 0.67 to 1.28 t ha^{-1} (8.8 per cent to 16.9 per cent). Generally, all tested VIs performed almost equally well at any crop growth stage. The same applies to tested classification thresholds r^2 - 4. Although some of the more extreme thresholds r^1 and r^5 showed satisfying results, these thresholds cannot be recommended, because of potential over- or under-estimation of crop coverage. The most suitable resolution and modeling strategy depends on the crop growth stage **(Figure 10.1)**. Due to row-based cultivation of corn and missing canopy closure, early growth stages require very high resolution imagery for accurate 3D crop surface models (CSMs) computation and classification-based separation of crop and soil. Compared to using simple unclassified mean crop heights (S1), prediction results significantly improve, when accounting for additional crop/ soil classification information (S2 and S3). With ongoing crop development and beginning canopy closure, high resolution imagery gets less and less important, sometimes even disadvantageous, due to higher noise. Good prediction results are achieved at intermediate resolutions by considering crop coverage as the second predictor variable (S3). With the completion of canopy closure, neither high resolution imagery nor crop/soil classification show potential to further improve prediction. Concluding these findings, combined spectral and spatial modeling, based on aerial images and CSMs, proves to be a suitable method for mid-season corn yield prediction.

10.4. Wheat and Barley

Jureèka *et al.* (2018) used remote sensing for yield estimation prior to harvest of set of crops (rape seed, wheat and barley) at the field level to enable agricultural decision making in Polkovice, located at low elevations in the Czech Republic. They used two datasets of satellite imagery namely Moderate Resolution Imaging Spectroradiometer (MODIS) and Landsat 8 datasets and compared with yields of ground truth. We tested a relationship between remote sensing data and the spatial yield variability measured by a yield monitor from a combine harvester. Correlations varied from approximately $r = 0.4$ to $r = 0.7$ with the highest correlation ($r = 0.74$) between yield and the Green Normalized Difference Vegetation Index

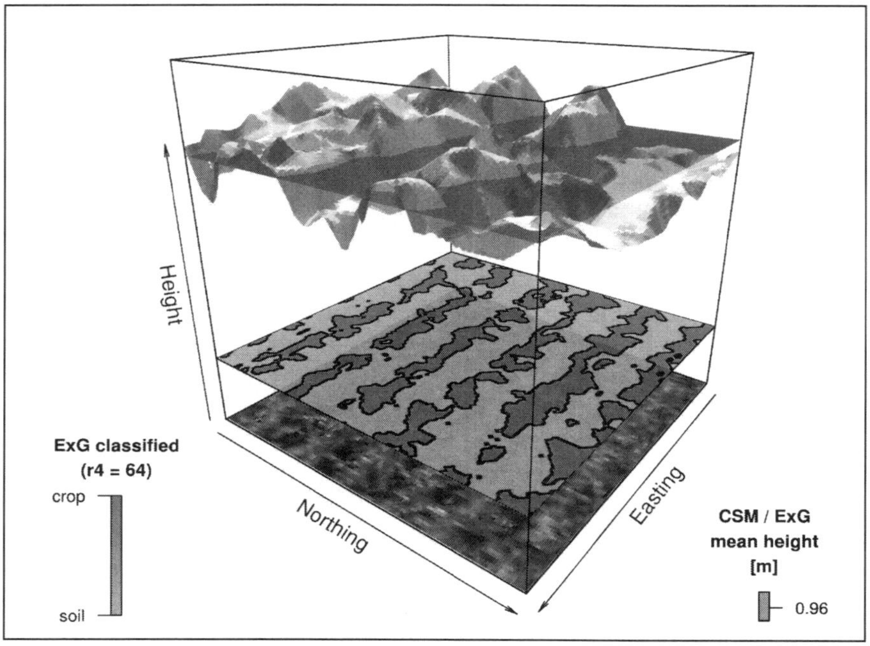

Figure 10.1. Mean Crop Height Computation using the Example of a 4 x 4 m Sub-sample of Plot 413 with a Sowing Density of 11 Seeds_m^{-2} and Nitrogen Application of 50 kg_N_ha^{-1}. The lower part of the figure shows a stack of the RGB orthoimage and the ExG layer classification based on threshold r4 at a ground resolution of 0.04 m and at crop growth stage Z39. The upper part shows the corresponding CSM layer height information as a 3D representation, colored by the ExG-classification. Mean crop height was calculated by the crop-classified CSM layer heights only and is displayed as a semi-transparent plane (*Source:* Geipel *et al.,* 2014).

collected from a drone. Vegetation indices from both Landsat 8 and the MODIS showed a positive relationship with yields for the compared period. The highest correlation was between yield and the Enhanced Vegetation Index (r = 0.8) while the lowest was between yield and the Normalized Difference Vegetation Index from MODIS (r = 0.1). The data have clearly shown that the utilization of remote sensing products can be useful for both the field as a whole and for its parts that can by visually assessed as homogenous, and some deviations or stress occurrence could be delimited using Vis (vegetative index) from remote sensing. As a non destructive method, remote sensing can be utilized several times during the growing season. It can provide crucial information about crop development and be used for real time agronomic decision making.

10.5. Wheat

Du and Noguchi (2017) exploited multi-temporal UAV orthomosaic images to visualize the rapid change of wheat growth status through image interpretation and discerned that the canopy greenness of the wheat field of study reached peak condition and the process of yellowing began thereafter. The imageries could bring out within-field variation of wheat tiller densities, especially from an early stage of wheat growth and occurrence of lodging was spotted in orthomosaic images taken. They reached the conclusion that orthomosaic UAV images taken at early stage of wheat growth could be used as prescription of variable-rate fertilizing to avoid, or alleviate, the occurrence of lodging by reducing dozes of fertilizer around over-high tiller density areas. The orthomosaic image taken ahead of harvesting could be used to generate navigation maps that guide either drivers of combine harvesters, or autonomous harvesting vehicles, to adjust operation speed according to the specific lodging situations. Through stepwise regression analysis of the response variable of sampled grain weight per square meter and the predictive variables of accumulative color vegetation indices, they concluded that only the variable of normalized green-red difference index was removed from the stepwise regression model due to insignificant p-value, whilst the rest variables of visible-band difference vegetation index (the normalized green-blue difference index, green-red ratio index, and excess green vegetation index) were included to fit the regression model, with coefficient of determination and RMSE as 0.94 and 0.02, respectively. The averaged value of sampled grain weight per square meter was 0.86 kg. The regression model was also validated by using leave-one-out cross validation method, which showed that the root-mean-square error of predication of the regression model was 0.06. Based on the stepwise regression model, a map of estimated grain weight per square meter (yield map) was generated by extracting each pixel values out of the maps of accumulative vegetation indices. Within-field spatial variations of wheat yield could be observed from the map, which could be seen as the comprehensive presentation of the spatial variations of soil fertility, tiller density, effective water potential, canopy aeration condition, and so on. They also obtained general information of the studied field that about 25.8 per cent areas in the studied field had grain weight per square meter below 0.5 kg, whilst grain weight per square meter of most areas reached between 0.5–1.5 kg, occupying about 50.4 per cent acreages. The mean value of grain weight per square meter was calculated as 0.72 kg, which indicated the estimated average yield of the studied field as 7.2 t/ha.

Fu *et al.* (2020) explored the potential of multi-spectral camera mounted on a multi-rotor UAV for monitoring wheat growth indices. The results showed that the vegetation index composed of a red edge band and near-infrared band was significantly correlated with LAI (leaf area index) and LDM (leaf dry matter). The optimal LAI estimation model was built by RESAVI (vegetation index combined with the red edge band and the near-infrared band), explaining 76 per cent variation, with an RRMSE (relative root mean square error) of 0.1990. The optimal LDM estimation model was built by NDRE (normalized difference red edge), explaining 74 per cent variation, with an RRMSE (relative root mean square error) of 0.2337.

Fu *et al.* (2020) also analyzed and evaluated the yield estimation accuracy of simple linear regression (LR), multiple linear regression (MLR), step-wise multiple linear regression (SMLR), partial least square regression (PLSR), artificial neural network (ANN), and random forest (RF). The results showed that the yield estimation model built by multiple factors is superior to the single factor, and the yield estimation model built by machine learning methods (PLSR, ANN, and RF) performed better than other models. In this experiment, the optimal yield estimation model was built by NDVI at the jointing stage, normalized difference vegetative index (NDVI) at the booting stage, NDVI at the flowering stage, and NDVI at the filling stage. The modeling method we used was RF, explaining 78 per cent variation, with RRMSE of 0.1030. In summary, this study demonstrates the potential of using a multi-rotor UAV combined with the multi-spectral camera to monitor wheat growth parameters and estimate yield. A variety of linear and nonlinear modeling methods were used to explore the possibility of further improving the accuracy of yield estimation. The UAV sensing system in this study can provide reference and technical support for the management and decision-making of intensive farming at medium-scale cropping area, which performed more efficient than handheld sensors. This method should combine with satellite data to enlarge the application at large-scale agricultural area in future study.

10.6. Barley

The UAV technology is widely used for data acquisition as it is relatively easy, timely, flexible, and cheap. This provides timely, detailed information on the current status of plants, which is valuable for breeders, insurance companies, and farmers. Breeding trials are particularly difficult to monitor on a regularly basis within a reasonable time, resulting in an increasing need for faster selection of superior lines. **Wilke *et al.* (2019)** have used UAV-based canopy height (CH) assessment that provides spatial information on the canopy height distribution and offers much more information compared to the classical plant height measurements of single spots in the field. The UAV-derived CH enables simplification of a complex crop surface with an objective scale to estimate genetic effects. The lodging assessment approach was done based on 3D canopy structure. As it is an independent measurement, this CH is very reasonable and practically feasible and relate growth and lodging behavior of the crop varieties. **Wilke *et al.* (2019)** have shown that areas affected by lodging could be detected with high accuracy even at the lowest spatial resolution (1.57 cm GSD). The higher the flight altitude is, the shorter the flight time, the smaller the number of recorded images, and the shorter the processing time. Therefore, fixed-wing UAVs, normally operated at higher altitudes to cover large areas, can be used for lodging assessment.

The permanent displacement of a plant from the upright position can be affected by different lodging severity, as illustrated in **Figure 10.2.** The first example shows a slightly affected canopy where only low-yield loss can be expected **(Figure 10.2a).** The same applies for the plot in **Figure 10.2b,** where just a few small areas are strongly affected by lodging. Additionally, **Figure 10.2c** clearly illustrates the capability of the approach to detect lateral lodging. In contrast to the other examples, the canopy in **Figure 10.2d** is heavily affected by lodging, so that parts of the plot lay completely on the ground.

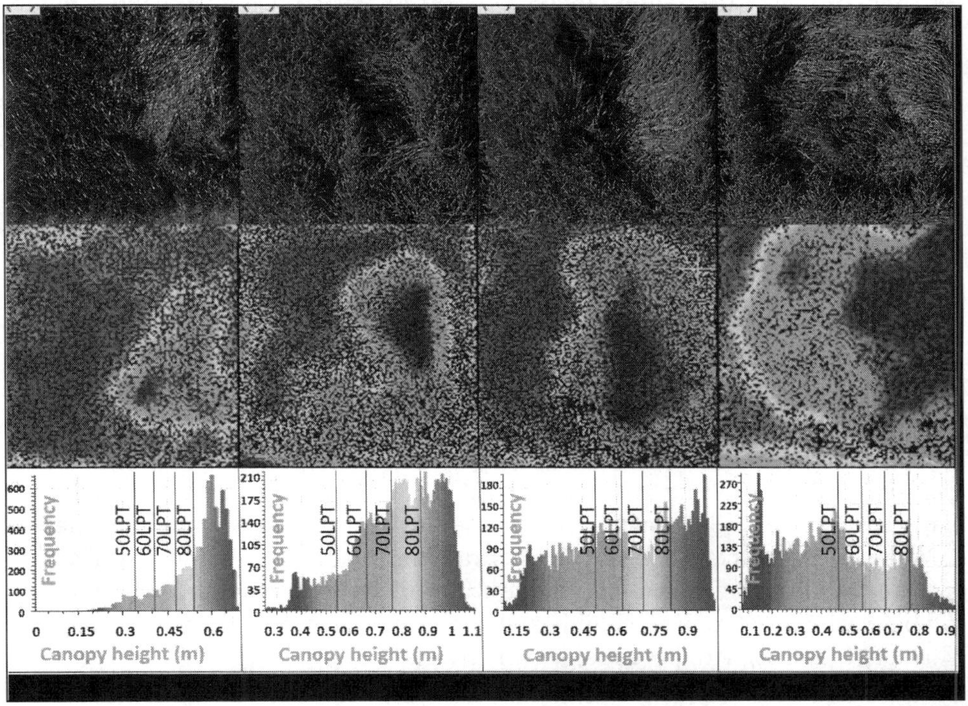

Figure 10.2. RGB Images of Barley Plots showing different Types (a–d) of Lodging (Top) and Corresponding Lodging Severity Derived from the CHM (Middle), as well as Canopy Height Distributions with Visualization of different Lodging Percentage Thresholds (LPTs) (Bottom) (Source: Wilke *et al.*, 2019).

Wilke *et al.* (2019) have used UAV-based approach to assess quantitative threshold in order to determine lodging percentage in a first step. The derived lodging estimates showed a very high correlation to reference data (R2 = 0.96, root mean square error (RMSE) = 7.66 per cent) when applied to breeding trials, which could also be confirmed under realistic farming conditions. As a second step, an approach was developed that allows the assessment of lodging severity, information that is important to estimate yield impairment, which also takes the intensity of lodging events into account. Both parameters were tested on three ground sample distances. The lowest spatial resolution acquired from the highest flight altitude (100 m) still led to high accuracy, which increases the practicability of the method for large areas. Their new lodging assessment procedure can be used for insurance applications, precision farming, and selecting for genetic lines with greater lodging resistance in breeding research.

10.7. Oats

Acorsi *et al.* (2019) employed a recent approach of introducing a model crop biomass consists in the use of RGB (red, green, blue) stereo images acquired from unmanned aerial vehicles (UAV) coupled with photogrammetric softwares to predict biomass through plant height (PHT) information. They generated prediction models for fresh (FBM) and dry biomass (DBM) of black oat crop based on multi-temporal UAV RGB imaging. Flight missions were carried during the growing season to obtain crop surface models (CSMs), with an additional flight before sowing to generate a digital terrain model (DTM). Models were established based on PHT derived from CSMs and field measurements, which were later used to build prediction maps of FBM and DBM. The study demonstrates that UAV RGB imaging can precisely estimate canopy height (R2 = 0.68–0.92, RMSE = 0.019–0.037 m) during the growing period. FBM and DBM models using PHT derived from UAV imaging yielded R2 values between 0.69 and 0.94 when analyzing each mission individually, with best results during the flowering stage (R2 = 0.92–0.94). Robust models using datasets from di_erent growth stages were built and tested using cross-validation, resulting in R2 values of 0.52 for FBM and 0.84 for DBM. Prediction maps of FBM and DBM yield were obtained using calibrated models applied to CSMs, resulting in a feasible way to illustrate the spatial and temporal variability of biomass. Altogether the results of the study demonstrate that UAV RGB imaging can be a useful tool to predict and explore the spatial and temporal variability of black oat biomass, with potential use in farms.

10.8. Soybean

Zhang *et al.* (2019) established plot-yield prediction models using unmanned aerial vehicle (UAV)-based hyperspectral remote sensing for yield-selection in large-scale soybean breeding programs. Three sets of soybean breeding lines were tested for plot yield and canopy spectral reflectance on 454~950 nm bands at different growth stages using a UAV-based hyperspectral spectrometer (Cubert UHD185 Firefly). In the establishment of plot-yield prediction models for soybean breeding programs using UAV-based hyperspectral remote sensing, four model construction elements were studied individually with the results being: (i) the suitable sampling unit-size in a plot was the central part of 20 per cent -80 per cent plot size (the high end was used in model construction to make a full use of the information); (ii) NDVI and RVI and their combination along with their best spectra combinations of near-infrared and red spectrum were the best vegetation indices for yield-prediction; (iii) seed filling stage was the best growth stage for a single-period model, while seed filling and seedling were the best combination for a two-period prediction model; (iv) linear regression was suitable for plot-yield prediction model construction in comparison to exponential and logarithm regression.

Seventeen prediction models composed of NDVI and RVI vegetation indices at seed filling growth stage and then 21 prediction models composed of the two vegetation indices at two growth stages (seed filling and vegetative) were established. The established model construction elements and methods could be used in the establishment of local models for pre-harvest yield-selection and post-harvest integrated yield-selection in advanced breeding nurseries as well as yield potential prediction in plant-derived-line nurseries, furthermore, these models can be used jointly for plot-yield prediction in soybean breeding programs.

10.9. Rapeseed

Gong *et al.* (2018) developed an approach to estimate rapeseed yield with remotely sensed canopy spectra and abundance data by spectral mixture analysis. A six-band image of the studied rapeseed plots was obtained by an unmanned aerial vehicle (UAV) system during the rapeseed flowering stage. Several widely used vegetation indices (VIs) were calculated from canopy reflectance derived from the UAV image. And the plot-level abundance of flower, leaf and soil, indicating the fraction of different components within the plot, was retrieved based on spectral mixture analysis on the six-band image and end member spectra collected in situ for different components. The data showed that for all tested indices VI multiplied by leaf-related abundance closely related to rapeseed yield. The product of Normalized Difference Vegetation Index and short-stalk-leaf abundance was the most accurate for estimating yield in rapeseed under different nitrogen treatments with the estimation errors below 13 per cent.This study gives an important indication that spectral mixture analysis needs to be considered when estimating yield by remotely sensed VI, especially for the image containing obviously spectral different components or for crops which have conspicuous flowers or fruits with significantly different spectra from their leaves.

10.10. Cotton

Feng *et al.* (2019) evaluated the performance of a UAV-based remote sensing system with a low-cost RGB camera to estimate cotton yield based on plant height. The UAV system acquired images at 50 m above ground level over a cotton field at the first flower growth stage. Waypoints and flight speed were selected to allow >70 per cent image overlap in both forward and side directions. Images were processed to develop a geo-referenced orthomosaic image and a digital elevation model (DEM) of the field that was used to extract plant height by calculating the difference in elevation between the crop canopy and bare soil surface. Twelve ground reference points with known height were deployed in the field to validate the UAV-based height measurement. Geo-referenced yield data were aligned to the plant height map based on GPS and image features. Correlation analysis between yield and plant height was conducted row-by-row with and without row registration. Pearson correlation coefficients between yield and plant height with row registration for all individual rows were in the range of 0.66 to 0.96 and were higher than those without row registration (0.54 to 0.95). A linear regression model using plant height was able to estimate yield with root mean square error of 550 kg ha-1 and mean absolute error of 420 kg ha-1. Locations with low yield were analyzed to identify the potential reasons, and it was found that water stress and coarse soil texture, as indicated by low soil apparent electricity conductivity (ECa), might contribute to the low yield. The findings indicate that the UAV-based remote sensing system equipped with a low-cost digital camera was potentially able to monitor plant growth status and estimate cotton yield with acceptable errors.

10.11. Sugarcane

Som-ard *et al.* (2018) introduced high spatial resolution UAV image for sugarcane yield estimation involving aerial images and ground data. The very high spatial resolution image along with object based image analysis (OBIA) showed great potential for estimating sugarcane yield before harvest. It predicted yield with more than 90 per cent of accuracy. Sugarcane industry and farmers can estimate sugarcane yield before harvest by utilizing the approach discussed in this study. The techniques discussed here have very less data requirement as compared to the traditional and conventional methods. The data requirements are only UAV images and ground survey data for height, weight, and number of stalks in each class of sugarcane. Again, UAV images have a higher potential for sugarcane yield estimation as the accuracy of yield estimation mostly depends on the accurate identification of sugarcane pixel area. Further study can be done to extract sugarcane height from UAV-based stereo images to reduce the ground-based observation effort. The structure from motion (SfM) technique can be used to generate the crop surface model (CSM). Then, it must be subtracted from the ground-based model to estimate the crop height (Bendig *et al.*, 2014; Holman *et al.*, 2016). Furthermore, the density of sugarcane can be estimated with a ratio of the leaf in particular area. The total area of the leaf can be extracted as the LAI index values by using the proportion of sky within crown envelopes. The sky view factor observed under the canopy from the zenith angle of the fisheye camera was successfully used to measure the crown cover and LAI (Pekin and Macfarlane 2009). Additionally, the total leaf area is also directly related to crop health and yield since the process of photosynthesis takes place in the leaves (Castro-Nava *et al.*, 2016). Therefore, it is possible to incorporate with the presented methodology for the continuous monitoring of crop health.

10.12. Alfalfa

Noland *et al.* (2018) used remote sensing technologies (*e.g.*, measurement of canopy reflectance) for rapid measurements of yield and nutritive value at the field scale estimation of alfalfa (*Medicago sativa* L.). The Canopy reflectance (350–2500 nm) and Light Detection and Ranging (LiDAR)-estimated plant height were measured to quantify the yield **(Figure 10.3).** Sets of specific spectral wavebands were determined via stepwise regression to predict alfalfa yield and nutritive value and models were reduced by spectral range to improve utility. They used Cumulative growing degree units (GDUs) and canopy height as model covariates. An alternative GDU calculation (GDUALT) using a temporally graduating base temperature was also tested against the traditional static base temperature. Their data have shown that the inclusion of GDUALT increased prediction accuracy for all response variables by 9–17 per cent. Models using a common set of seven wavebands, combined with GDUALT, explained 81–90 per cent of the variability in yield, crude protein (CP), neutral detergent fiber (NDF), and NDF digestibility (NDFd; 48-h invitro), respectively.

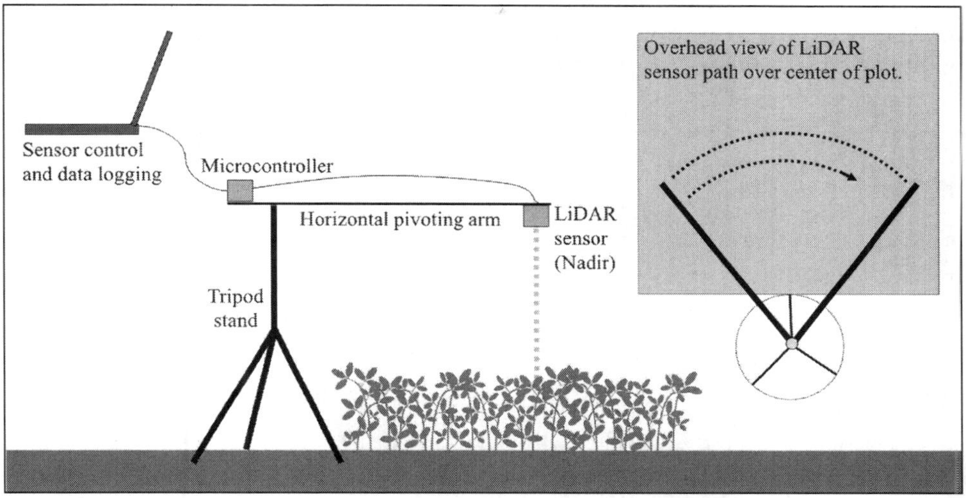

Figure 10.3. Illustration of LiDAR Sensor Deployment to Estimate Canopy Height in Alfalfa Plots (Source: Noland *et al.*, 2018).

Noland *et al.* (2018) established the potential for remote sensing measurements integrated with air temperature information to achieve rapid and accurate predictions of alfalfa yield and nutritive value at the field scale for optimized harvest management. This work establishes potential for integrating environmental data such as GDUs with canopy reflectance data and remote measurements of canopy height for improved estimations of alfalfa yield and nutritive value. These methods could be applied on-farm across large acreages and multiple fields to inform economically efficient decisions regarding timing of harvest, which field(s) to harvest first, or how to ration the current feed supply relative to the anticipated harvest time, yield, and nutritive value. Modern software applications can easily retrieve environmental data such as estimated soil type, temperature, rainfall, and relative humidity based on geographic location, and could apply iterative regression with every harvest to improve predictive models.

10.13. Wild Tomato

Johansen *et al.* (2019) used remote sensing tool to predict the wild tomato species, *Solanum pimpinellifolium*, through field and UAV-based assessment. They used Random Forest approach to model biomass and yield. The results showed that shape features such as plant area, border length, width and length had the highest importance in the random forest models. A week prior to harvest, the explained variance of fresh shoot mass, number of fruits and yield mass were 86.60 per cent, 59.46 per cent and 61.09 per cent, respectively **(Figure 10.4).** The explained variance was reduced as a function of time to harvest. Separate models may be required for predicting yield of salt-stressed plants, whereas the prediction of yield for control plants was less affected if the model included salt-stressed plants. This research demonstrates that it is possible to predict biomass and yield of tomato plants up to four weeks prior to harvest, and potentially earlier in the absence of severe weather events.

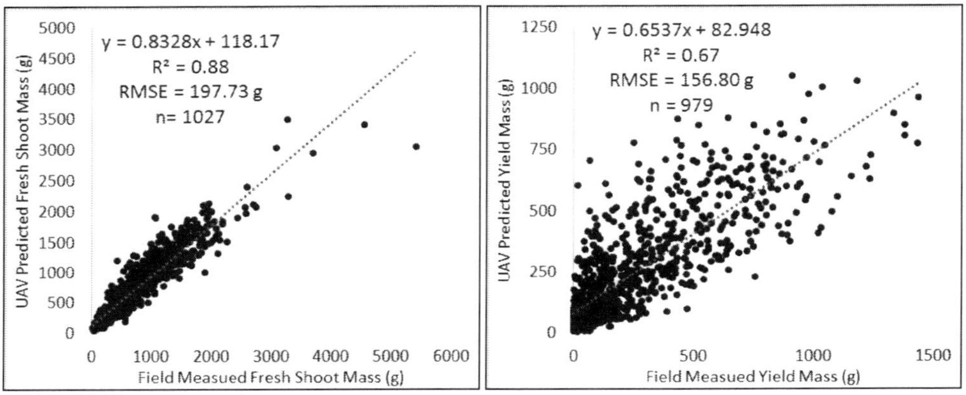

Figure 10.4. Scatterplots Showing the Correlation between Field and UAV Measured (a) Fresh Shoot Mass and (b) Yield Mass Based on UAV Imagery Collected on 7 January 2018 (Source: Johansen *et al.,* 2019).

Overall, the literature review suggests that UAVs are highly useful, reliable and rapid in the assessment of crop growth and yields well before the harvest of the crop. This is very important for the farmers to know the crop yields at the maturity phase in order to adjudge the income. This is also helpful for the insurance companies to make judgment of the crop gain or loss. The UAV studies conducted for various crops to determine the yields of the standing crop before harvest are summarized in Table 6.

- ✰ The UAVs have been experimented to assess the yields of several crops including, rice, wheat, maize, oat, soybean, cotton, sugarcane and tomato. Since these crops are cultivated in larger extent and the price of the produce is fluctuating, aerial survey and imageries will be helpful to assess the marketing potential well before the crops are harvested.
- ✰ In order to accomplish the task of yield assessment, RGB, digital camera, multi-spectral or hypersepectral camera are widely used as sensors. These imageries are converted into valuable and quantifiable data using logarithms.

Table 6. Application of UAV in Crop Yield Assessment

Sl.No.	*Crops*	*Sensors*	*UAV Type*	*Model Input*	*Model Output*	*Model type*	*Reference*
1	Rice	Digital cameras	UAV	Vegetation index/ spectral mixture analysis (SMA)	Estimate yield	Linear Spectral Mixture Analysis	Duan *et al.*, 2019
2	Rice	-	UAV	NDVI	Estimation of yield	Solar radiation datasets.	Akira *et al.*, 2017
3	Wheat	Digital camera	UAV	color vegetation indices (CVI)	Wheat yield map	Regression model	Mengmeng *et al.*, 2017
4	Wheat	Multi-spectral Camera	Multi-rotor UAV	NDVI	Wheat yield estimation	Simple Linear regression (LR), multiple linear regression (MLR), stepwise multiple linear regression (SMLR), Partial least squares regression (PLSR), artificial neural network (ANN), and random forest (RF) Modeling methods	Zhaopeng *et al.*, 2020
5	Barley	RGB	UAV	pixels	Assessment of lodging severity	Structure from Motion (sfm) techniques	Norman *et al.*, 2019
6	Black oat	RGB	UAV	Plant height (PHT) information	Estimating biomass	Crop surface models (csms	Matheus *et al.*, 2019
7	Maize	RGB	UAV	Pixels	Crop yield prediction	3d crop surface Models (csms)	Jakob *et al.*, 2014
8	Maize	RGB	UAV	GNDVI/SPAD readings	Prediction of crop vigor and yield	Regression model	Ibrahim *et al.*, 2018
9	Cotton	RGB	UAV	Plant height/Digital Elevation Model (DEM)	Cotton yield estimation	Linear regression model	Feng *et al.*, 2019
10	Soybean	Hyperspectral Spectrometer	UAV	NDVI	Yield evaluation	Multiple regression	Xiaoyan *et al.*, 2019

Sl.No.	*Crops*	*Sensors*	*UAV Type*	*Model Input*	*Model Output*	*Model type*	*Reference*
11	Rapeseed	Customer-Specified Band	UAV	Pixel, NDVI	Yield estimation	Spectral mixture analysis	Yan Gong *et al.*, 2018
12	Sugarcane	RGB	UAV	Object-based image analysis (OBIA) and and pixel-based image analysis	Sugarcane area and yield estimation	Feature Envelope to Polygon" tool in arcgis 10.2	Jaturong *et al.*, 2018
13	Sugarcane	RGB and multispectral images	UAV	UAV-derived LAI and soil water content (SWC)	Sugarcane yield estimation	Crop growth model (e.g. SWAP-WOFOST)	Liangsheng *et al.*, 2018
14	Tomato	RGB	Quad-copter	Pixels/object-based plantdelineation	Prediction of Biomass and yield	Random forest	Johansen *et al.*, 2019

- ☆ UAVs, multi-rotor UAVs and Quadcopter are used as aerial vehicles to assess the yield of crops. UAVs are simple, non-destructive and repeated values can be obtained to gain reproducible data.
- ☆ The indices used for the harvesting time and yield assessment of standing crop encompass vegetative index, NDVI, GNDVI, OBIA (object based image analysis) besides indirectly through LAI, plant height and soil water content. Among the various indices evaluated, NDVI is widely used and established a robust and strong correlation with yield of crops. In all the cases, yield was referred as a dependent variable.
- ☆ Statistical models exploited for the crop yield assessment using UAVs include linear regression, multiple-linear regression, step-wise multiple linear regression, Partial least squares regression (PLSR), artificial neural network (ANN), and random forest (RF). Linear regression model is widely for the yield assessment.

This chapter gives overview of the use of UAVs on yield assessment using various qualitative and quantitative parameters. The dada strongly suggest that UAVs are very effective, cheap and non-destructive way of quantifying the crop yields prior to harvest. This will help the farmers to get lucrative price when they harvest.

Chapter 11

Application of Drones in Biodiversity

Fast growing population, urbanization and expansion of human habitation and industrialization threaten biodiversity of plants, animals, microbes and all creatures on earth. The wildlife populations are undergoing alarming declines, and extinction rates are now as high as 100-fold greater than the background extinction rate (Ceballos *et al.,* 2015). Ecological monitoring is essential for understanding these population dynamics, and rigorous monitoring facilitates informed management. The effectiveness of management decision-making is often dependent on the accuracy and timeliness of the relevant ecological data upon which decisions are based, meaning that improvements to data collection methods may herald improved ecological outcomes from management actions. Emerging technologies are increasingly being adopted by ecologists to improve data collection and capture novel data (Pimm *et al.,* 2015). Advances in genetic techniques have resulted in the cost-effective application of environmental DNA sampling to the detection of endangered species and invasive species (van Rooyen, and McCarthy, 2015; Smart *et al.,* 2016). Camera traps and acoustic recorders have become established tools for determining whether a species is present at a site, and estimating population density (Marques *et al.,* 2013; Rowcliffe and Carbone, 2008). Furthermore, animal-borne telemetry devices have revolutionized our understanding of animal movements, including their interactions with the environment, and species distributions (Hussey *et al.,* 2015; Kays, Crofoot, Jetz, and Wikelski, 2015). Such technologies have been vital in advancing our understanding of wildlife and answering fundamental questions, such as how many individuals are in a population and whether that population trajectory is increasing or decreasing.

Drones have seen a rapid uptake by ecologists for data collection. This surge in popularity has arisen largely due to their ability to carry remote sensing instruments that collect data at scales highly suited to monitoring ecological phenomena (Anderson and Gaston, 2013). Compared to remote sensing instruments mounted to spacecraft and conventional aircraft, drones are more suited to collecting extremely fine spatial and temporal resolution data at the discretion of the user. These benefits have led many practitioners to label drones as a powerful tool for wildlife ecology (Lejeune and Vermeulen, 2015). Consequently, drones are being used to transect counts of African elephants *Loxodonta africana* (Vermeulen, Lejeune, Lisein, Sawadogo, and Bouche, 2013), monitoring for poaching activities (Mulero-Pazmany, Stolper, van Essen, Negro, and Sassen, 2014), detecting reptile and arboreal mammal nests (Evans, Jones, Pang, Saimin, and Goossens, 2016) and many bird species hooded crow *Corvus cornix* (Weissensteiner, Poelstra, and Wolf, 2015; Wilson, Barr, and Zagorski, 2017). The Robotic Process Automation (RPA) -derived imagery would be more accurate and more precise than those generated using the traditional approach, confirming that RPA technology is a significant advance for ecological monitoring.

11.1. Elephants

Addressing human–wildlife conflicts is a fundamental challenge for conservation practitioners. In some areas of the planet, the loss of lives, crops, or live-stock because of wildlife has significant consequences for people's livelihoods and their food and agricultural security. Reconciling the conservation of endangered species with human activities, as well as with socioeconomic uses and human security, has been the subject of numerous studies and management programs. The most commonly implemented actions are aimed at mitigating wildlife negative effects on agricultural production and human health through mitigation operations or economical compensations for losses caused by wildlife. Quantifying and mapping wildlife-caused damages is essential to carry out both kinds of actions. Most methodologies are aimed at developing predictive risk maps, but in the context of human–wildlife conflicts, it is crucial to develop accurate protocols for the reliable verification of the authority of the causative species and their relation with damage claims. These protocols are fundamental in creating public trust in the legitimacy of compensation programs, and in avoiding fraud and moral hazards. Finding simple and inexpensive methods to quantify damages is a major challenge in decreasing the cost–benefit ratio to achieve conservation objectives. Medium resolution remote sensing imagery as low-cost Unmanned Aerial Vehicles (UAV) has arisen as an essential tool to meet this challenge.

Man-animal conflict particularly humans and elephants is major problem in many of the states in India. The States that juxtaposed with maintains or farm lands adjacent to the reserved forests are frequently disturbed by elephants. Crop raiding is one of the most common and costly forms of conflict between people and elephants, provoking both retaliatory killing of elephants, and animosity towards elephants and other wildlife among local communities. The latest statistics suggests that India has 29,964 elephants distributed across the country and major States being Karnatka (6049), Assam (5719), Kerala (5706), Tamil Nadu (2761) and other States. About 44 per cent of the elephant population lives in Southern India Karnataka, Kerala and Tamil Nadu. During 2015-18, human elephant conflict caused 1713 human deaths, 373 elephant deaths due to unnatural cases, including poaching and electrocution **(Figure 11.1).** The forest officials say that urbanization, habitat disturbance, frequent drought and restricted water availability in the jungle

Figure 11.1. Elephant-Human Conflict: A Cause for Concern.

are the main reasons for the elephants to visit farm lands causing damage to the cultivated crops like maize, sugarcane, banana and bamboo.

Farmers employ a range of measures to reduce crop raiding, including fire, drums, electric fences, early warning systems, and Capsicum (chilli) and beehive fencing, electric fencing, fire and crackers. Among them crackers and fire makers are frequently and commonly used as deterrence These methods often require close proximity to elephants, increasing the against elephants. This causes a chance of injury or death to people and elephants because frightened elephants can be aggressive. Actively defending crops also disrupts daily routines and school schedules, as farmers and their children must remain vigilant throughout the night during the harvest season.

In order overcome human-elephant conflicts, studies have been conducted to employ unmanned aerial vehicles (VAVs) to monitor and repel them from the human habitat. Hahn *et al.* (2016) have shown that elephants reacting swiftly and dramatically to drones being used for aerial photography. The drones had blinking red and green lights and emitted a loud whirring noise have scared away the elephants from cultivated areas. These observations prompted us to test the efficacy of drones operated by trained wildlife managers to move elephants out of conflict zones.

Hahn *et al.* (2016) employed four drone model namely Phantom 1, Phantom FC 40, Phantom 2 Vision, and Phantom 2 Vision+ (DJI, Shenzhen, China) and found that all models produce similar whirring sounds and have identical red and green blinking lights that deter elephants. They indicated that there is a consistent success of drones as a mitigation tool during three crop seasons and among different elephant populations. African Farmers suggest that drones have the potential to be an effective, long-term and cost-effective tool for human–elephant conflict mitigation. Their success during all of the night trials is particularly encouraging, as the dusk-to-dawn period is a notoriously challenging time to move elephants. Drones are also easy to bring in and out of operation as crops ripen, and can be shifted rapidly to areas of intensifying conflict or where elephants enter populated areas. Deploying drones for human–elephant conflict mitigation can also reduce the need for permanent barriers, such as electric fencing, which are broken regularly. Nonetheless, their data suggest that drone-based methods offer a promising new tool to integrate within a broader suite of proactive, community-based, low-cost mitigation approaches to reduce negative interactions between people and elephants.

Drones have been used to detect the animal population in wild. Drones were successfully employed to enumerate animals in wild habitat such as elephants, tigers, rhinoceros, crocodile, hippopotamus, chimpanzee and other animals. The details of use of drones in the assessment of biodiversity in Africa and Asia are furnished in **Table 7.**

Table 7. Utilization of UAV in Wildlife Monitoring in Africa and Asia

Sl. No.	*Country*	*Species*	*Monitoring*	*UAV type*	*Camera Sensors*	*Flight Height*	*Reference*
1	Sumatra, Indonesia, Asia	*Sumatran orangutan* and Sumatran elephant	Detection	Hobbyking Bixler	Cannon IXUS 220 HS (Pentax) and GoPro Optio WG-1 GPS	100 m, 180 m	Koh *et al.*, 2012
2	Nepal, Asia	Endangered species: Tiger	Endangered species against poachers	GPS-enabled FPV Raptors	Still and video cameras	<200 m	Goodyer *et al.*, 2013
3	Namibia, Africa	Large mammals	identification	Canon PowerShot S110 compact camera,	Red, Green and Blue (RGB) bands	-	Rey *et al.*, 2017
4	Bukina Faso, Africa	African Elephant (*Loxodonta africana*)	Animal counts	Gatewing6100TM	Ricoh GR III camera	100 m and300 m	Vermeulen *et al.*, 2013
5	South Africa	Black rhinoceros (*Diceros bicornis*) and the white rhinoceros (*Ceratotherium simum*)	Rhinoceros and poachers detection	Fixed-wing, Easy Fly St-330 (St-models, China)	(i) FPV video camera, (ii) Panasonic Lumix LX-3 digital camera (iii) thermal video (IR camera module)	32–180 m	Mulero *et al.*, 2015; Rey *et al.*, 2017
6	Democratic republic of Congo, Africa	Common Hippopotamus (*Hippopotamus amphibius* L)	Population	Fixed-wing, Falcon UAS	Sony Nex7 digital still camera (RGB)	20 m, 40–150 m	Linchant *et al.*, 2015
7	South Africa	Nile crocodile (*Crocodylus niloticus*)	Population 287 were identified but only 211 crocodiles were Counted in the diurnal ground survey	DJI Phantom 3 Standard Drone	Still camera of 1/2.3" sensor, fast f/2.8 prime lens	40 m, 55 m and 70 m	Ezat *et al.*, 2018

Sl. No.	Country	Species	Monitoring	UAV type	Camera Sensors	Flight Height	Reference
8	Tanzania, Africa	Chimpanzee (Pan troglodytes)	Nests detection and population estimate	(i) Skywalker X5 frame; (ii) HBS Skywalker 100 km Long Range Fix Wings drone	(i) Canon S100 camera operating a CHDK firmware modification; (ii) Sony RX100M2	90m	Bonnin, *et al.*, 2018
9	Tanzania, Africa	African Elephant (*Loxodonta africana*)	Human-Elephant Conflict control	DJI Phantom	2 No camera	50 m	Hahn *et al.*, 2017
10	Tanzania, Africa	Human Poachers	Detection of poachers	Multicopter (DJI F550 frame)	(i) 16 MP RGB Survey 2 camera; (ii) Thermal Capture v1.0 TIR camera	70 m, and 100 m	Hambrecht *et al.*, 2019
11	Tanzania, Africa	Wetland habitats	Wetland condition and human threats	Multicopter DJI Phantom 4 Pro	20MP camera with 1 in sensor CMOS with a Field of View of 84.4 and a lens of 8.8 mm equivalent to a 24 mm	-	Bregoli, *et al.*, 2019

The Table 1 summarizes the following details and provides how best the drone technology can be used to detect the animal population, intensity and develop strategies to conserve them. Indeed, drone technology can also be used for detecting the poachers and develop suitable measures to contain the illegal activities in the areas could not be reached by humans.

- The Asian elephants can be detected using "**Hobbyking Bixler**" type of drone with a flying height of 100-200m with a suitable sensing device like sensors. On the other hand, African elephants can be enumerated using a flying height of 300 m.
- Tigers can be enumerated using **GPS enabled FPV Raptors** attached with high resolution cameras with a flying height of 200 m. Such assessment were used for enumeration and the intensity of poaching of wild cats
- Rhinoceros (white and black) were assessed used **fixed wing drones** attached with a FPV Video Camera with a flying height of 50-180 m.
- Drones were also used for the population enumeration for Hippopotamus using **Fixed wing type** attached with digital still camera. The flying height of the drones was in the range of 20-50 m.
- Crocodile population was detected using **DJI Phantom drone** affixed with still camera and the flying height of 40-50 m. Drones were used for both detection and enumeration of crocodiles.
- Chimpanzee population was counted using **skywalker type** of unmanned aerial vehicles attached with Canon camera with a flying height of 90 m.
- Apart from wildlife assessment and enumeration, drones can also be used for poaching activity in wild. **Multi-copter type** of drones is widely used for detecting the poaching which is very much important for the conservation of biodiversity.

The literature review suggests that drone technology is very appropriate to detect, enumerate and manage animal biodiversity in wild where humans direct inventions are hardly possible.

11.2. Waterbird

Waterbird communities are potential indicators of ecological changes in threatened wetland ecosystems and consequently, a potential object of ecological monitoring programs. Waterbirds often breed in largely inaccessible colonies in flooded habitats, **Afan *et al.* (2018)** have used unmanned aerial vehicle (UAV) surveys to provide a robust method for estimating their breeding population size. Counts of breeding pairs might be carried out by manual and automated detection routines. In this study we surveyed the main breeding colony of Glossy ibis (*Plegadis falcinellus*) at the Doñana National Park **(Figure 11.2).** They obtained a high resolution image, in which the number and location of nests were determined manually through visual interpretation by an expert. They also suggested a standardized methodology for nest counts that would be repeatable across time for long-term monitoring censuses, through a supervised classification based primarily on the spectral properties of the image and a subsequent automatic size and form based count. Although manual and automatic count were largely similar in the total number of nests, accuracy between both methodologies was only 46.37 per cent, with higher variability in shallow areas free of emergent vegetation than in areas dominated by tall macrophytes. They discussed the potential challenges for automatic counts in highly complex images.

Figure 11.2: (a) Glossy ibis (*Plegadis falcinellus*) individual. (b) Front view of the Glossy ibis colony, densely covered by the macrophyte Typha, which makes it extremely difficult ground census of the colony. Photos: Carlos Gutiérrez-Expósito (a), and Manuel Máñez (b).

11.3. Orthomosaic Images

The orthomosaic obtained from the aerial survey showed an area of 16 ha covering the complete surface of the Lucio de la FAO where Glossy ibis breeds **(Figure 11.3).** The Glossy ibis colony was based in a continuous area of 3.2 ha with two major classes of vegetation cover: flooded areas covered by macrophytes (mainly Typha) and free areas of emergent vegetation but covered with aquatic plants (*Lemna* spp). Regardless the vegetation type cover around the nests, Typha invariably constitute the predominant nesting material. No nests were found over Tamarix shrubs.

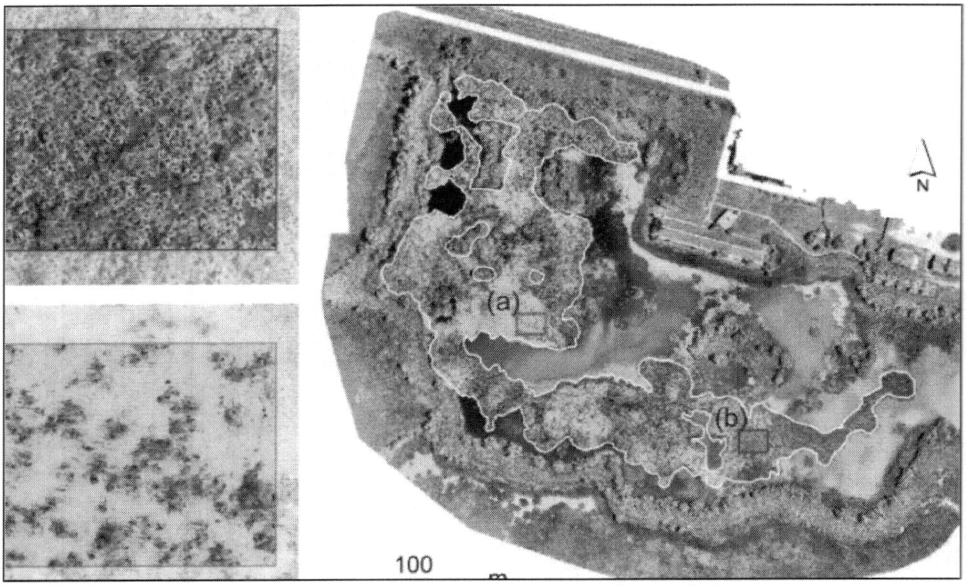

Figure 11.3: Orthomosaic Image of High-resolution of the Glossy Ibis Colony in "Lucio de la FAO". Image was mosaicked with Pix4D mapper Pro (Pix4D, Switzerland) after acquiring 448 photos through automatic flight of Phantom 4 Pro quadcopter (DJI Innovations, Shenzhen, China) on 14 May 2018. Squares show validation areas to test the accuracy of the automatic nest counting method to the visually nest count by an expert. Blue squares show densely macrophytes area (Typha spp) whereas red squares showfree areas of emergent vegetation covered by Lemna aquatic plant. White line delineates the Glossy ibis nesting area.

The literature review on drone technology in biodiversity of animals, birds, and others clearly demonstrated that unmanned vehicles are very appropriate in enumeration of wild animals where human interference or involvement is very hard and inaccessible. Despite the fact that the data base assists in understanding the biodiversity, there are lots to be done in the context of Indian perspective. This book provides a broad spectrum usage of drones and opens up various avenues to explore and exand its application.

Chapter 12

Bottlenecks in Drone Technology

Drone technology is very powerful and it is going to be employed in a wide spectrum of activities in digital agriculture. Since drones do the job quickly and rapidly, the response time is very short to save the crop from damages caused by pests and diseases. The country like India where more than 800 million people still depend on agriculture and facing acute labour shortage, certainly, drone technology will help the cover the area must faster and cheaper than human labours. The technology is fast growing in the country. It is appropriate to evolve suitable strategies to gain the advantages while circumventing the constraints. There are number of problems faced by the farmers are listed below that are to be addressed to augment the adoption rate.

12.1. Technical Decisions

There are mushrooming of drones manufacturers in the country and people are confused with the specifications, costs and its utility. In the market, there are many manufacturers and vendors for various UAVs from industrial products to hobby-type model aircraft. Due to the **lack of standard protocols** of UAV development for agricultural applications, it is difficult to decide which UAV is appropriate for any single project both technically and economically. Currently, there is no specific recommendation for the use of drone in agriculture either from the State or Central Governments. In Tamil Nadu, being the progressive State, it used drone technology for a massive spray against an invasive and a dreadful pest "Fall Army Worm (*Spodoptera furgiperda*)" in maize crop. While using the technology, there were number of problems faced by the users such as drifting, droplet size, concentration and effectiveness of the spray. All these need to be addressed in order to optimize specifications of drones to deliver the pesticide spray effectively. The literature reported in the scientific publications will serve as the framework to set the base and can be fine tuned to suit Indian conditions.

12.2 Cost

As farmers or public have started using drone technology for various purposes that include patrolling people during COVID 19 pandemic. Now, there are lots of manufacturers of drones in the country with varying cost. The industrial systems, especially spray UAVs, are typically expensive for individual farmers. Despite the fact that individual unit cost may be higher but custom hiring will substantially reduce the cost while improving the adaptability of the technology. In hobby shops, inexpensive airframes can be purchased and assembled with engine and avionics components. A light-weight, low-cost camera can be integrated into this assembly to typically build up a short endurance UAV LARS platform. Although the costs of the aircraft and the camera could be minimized, the assembly and integration require significant labor and time even for highly skilled technicians and engineers. Therefore, the cost of the technology may appear to be expensive at the beginning and the cost gets reduced when adoption rate increases which is supported by the government support and regulatory framework in place.

12.3 Payload

Payload size and weight are critical factors for UAVs. Payloads are of varying capacity depending on the types of spray and extent of spray. When deciding on an appropriate UAV for a specific application, the designated payload needs to be carefully configured to determine if UAV structure needs to be modified or optimized to accommodate it. When the UAV is ready, mechanical and electrical systems need to be configured. Payload design, mechanical, and electrical accommodation for UAVs have no general engineering guidelines to follow. However, payloads vary with type of application. In case of pesticides, the existing pesticide formulations available in the market is used and modulated to suit the available drones. Indeed, ultra-low volume spray formulation is preferred for drone technology and lots to be researched to optimize the payload.

12.4 Operation

One of the major issues to be addressed in drone technology is the skilled manpower. It is operated primarily by engineering graduates with experience in aeronautics and they may not know about the crops and their requirement. On the other hand, crop scientists would not be aware of drone operations. In this context, interfacing between agri-professionals and aeronautics engineers are utmost essential for the success of the technology. In some cases, automatic operation is available but it may be risky and jeopardize the livelihood of people. Most UAVs do not have the capability of automated take-off and landing. Therefore, skilled operators for RC control are required. Autonomous flight based on geo-referenced coordinates is therefore a highly desirable component for practical use of UAVs in agriculture. This is one of the emerging areas of interest in digital agriculture and needs lots of integrated skilled manpower.

12.5 Reliability

The UAVs are **sensitive** to other network communications. Therefore, UAVs are preferably operated in areas without wireless networks. Cellular phones and PDAs nearby may interfere with the communication between the ground station and the UAV in flight. The UAVs would have been built for surveillance and patrolling the public or scenic photography. Use of UAVs is very limited and beginning to scratch the surface in agriculture. The **mechanical structure** of many UAVs was not manufactured to be strong enough for agricultural operations. It is thus imperative that durability be assessed, but this is often difficult to do prior to purchase or setup of the UAV system. Our experience with a UAV helicopter configured for low volume spraying was a failure of the landing gear structure, resulting in damage to the rotors, landing gear, and servos on the rotor head. The repairs were costly in labor, time, and materials requiring special parts and knowledge of servo operation for proper repair. This significantly delayed the research project and confidence in use of UAVs for spraying was lost by the agricultural research developers and potential users. Yet another issue in drone technology is electronic failure. A supposedly 'fail-safe' electronic guidance system malfunctioned, causing a UAV to crash. The electrical UAV helicopter being used for remote sensing crashed from an altitude at about 7.6 m when communication was suddenly lost due to the electronic failure on the UAV. The UAV was supposed to hover above takeoff zone if the communication was lost but the failsafe function also failed. The crash resulted in a bent tail rotor assembly.

Chapter 13

Regulatory Guidelines for Drones Technology

Drone technology is gaining popular in agriculture primarily to overcome the acute labour shortage. While the drone technology gains momentum, one of the main barriers to rapid full-scale growth of commercial unmanned aircraft is the concern for safety. As myriad certification agencies scramble to keep up with the unique demands of this fast-growing industry, one thing is clear – where applicable, pertinent certification standards for manned aircraft are starting to apply. For the complex electronics that provide communication and control of these systems, this means a swift move towards compliance with **DO-178C** and **DO-254** for software and hardware development. In most cases, the unmanned aircraft can only be operated as part of a system, hence the term "unmanned aircraft system" or **UAS.** The UAS consists of an unmanned aircraft (UA), a remote pilot station and the command, control and communications links that join them; as such, safety considerations address all of these elements

In 2011, the **International Civil Aviation Organization** of the United Nations published Circular 328 – this document states a UAS should demonstrate equivalent levels of safety as manned aircraft and thus meet relevant government rules for flight and flight equipment. Within the United States, the Congress passed a bill in 2012 that mandated the Federal Aviation Administration (FAA) create a plan for allowing UAS into commercial airspace. Subsequently, the FAA issued "the Integration of Civil Unmanned Aircraft Systems (UAS) in the National Airspace System (NAS) Roadmap.

In 2014, USA made in mandatory that the Unmanned Aircraft System (UAS) requires an Experimental Worthiness Certificate to access National Airspace System. The FAA Order (8130.34) Air Worthiness Certification of Unmanned Aircraft System establishes procedures for issuing this certification, and as such establishes guidance standards for certification aspects of development and operation, which may be addressed by adoption of such standards as **DO-178C.**The FAA roadmap is, in essence, maturation of the acceptance of UAVs from this "experimental" aircraft certification to requiring the same standard airworthiness type certification of manufacturing design as is now required of conventional manned aircraft.

The **European Union** sees benefits and challenges for civilian drones, and in 2014 proposes a set of regulations to control the effects of drones on peoples' safety, security and privacy. The UAS Advisory Group was set up in 2015 by the United Nations' civil aviation arm to draw up global rules and regulations for the safe use of unmanned aircraft. The team comprises countries such as the United States, France and China, as well as industry bodies like the global pilots' association. In December 2017, the International Air Transport Association (IATA), a global airline trade body, pressed governments to ensure enforcement of regulations to curb reckless and dangerous flying of recreational drones. The regulations for drones is highly variable with countries but the regulatory framework serves as a build for our own for implementation in the country.

Australia

The Australian Government set a rule that the remotely operated aircrafts should be kept at least 30m away from structures, buildings and people accordingly to the Civil Aviation Safety Authority (CASA). Further, the operators of UAS should check with the local council where they could be used

Brazil

The National Civil Aviation Agency regulated the operation of drones through the Brazilian Special Civil Aviation Regulation No. 94/2017 (RBAC-E No. 94/2017) in 2017. ANAC's regulation complements the drone operating rules established by the Airspace Control Department (DECEA) and the National Telecommunications Agency (ANATEL).

Canada

The Transport Canada proposed the implementation of new regulations in 2016 that would require all drones over 250 grams to be registered and insured and that operators would be required to be a minimum age and pass an exam in order to get a license. These regulations were introduced in 2019.

France

Overflights of nuclear power plants are illegal in France, with a punishment of a year in prison and a fine of €75,000 if an aircraft comes within 5 km horizontally or 1 km vertically of a plant.

Hong Kong, China

While no-fly zones have been declared in Victoria Harbour, country parks, the airport, military sites, and prisons and in government-run leisure facilities, it was reported that rules were rarely enforced or observed. The Hong Kong government is proposing to introduce regulations that will broadly follow American standards, which might require all drones above a certain weight to be registered. Information would then made available to law enforcement authorities, and failure to register and tag a drone could be punishable by a US$250,000 (HK$1.95 million) fine or up to three years in jail. A short public consultation on the matter is set to be conducted by the first quarter of 2018, and to finish by mid-2018.

India

Permission from WPC (Wireless Planning and Communication) is required for importing any radio controlled equipments in India, including drones/UAV. Drones imported into India, without prior permission from WPC and DGCA, will be confiscated by Customs at point of entry. Drones above 2 kg also requires Air Defense clearance (Flight permit); and drone pilot who is least 18 years old, and has completed the 'drone pilot training'.

The DGCA (Directorate General of Civil Aviation) has on August 27, 2018, issued finalized drone guidelines. The regulations come into force from December 1, 2018. Under the regulations, drones are restricted items and cannot be carried in hand baggage. All drone operations are restricted to daylight and within visual line of sight. However, shooting in well-lit enclosed premises using micro drone up to 200 ft is permitted even after sunset.

The ICAR has developed guidelines for the use of drones in agriculture and the report is submitted to Ministry of Agriculture and Farmers in January 2020. Once it is approved by Government of India, the regulations will be in place for effective use of drone technology for various spheres of agriculture.

Indonesia

In 2015, the Indonesian Ministry of Transportation's Directorate General of Civil Aviation published a rule that regulates the usage of Unmanned Aerial Vehicles in the Indonesian Airspace. The rule states that UAVs should not fly above the altitude of 150m and should not fly inside restricted or prohibited area, and areas around airports. UAVs that require altitudes higher than 150m would require a written authorization from the Directorate General of Civil Aviation. Additionally, UAVs equipped with imaging equipment should not fly within 500m from the border of restricted or prohibited area, and if the UAV is involved in imaging activities, the operator should have written permission from the local government. Special UAVs equipped with farming equipment such as seed spreader or insecticide spray should only operate in farmland, and should not operate within 500m of housing area.

Instead of NFZ (No-fly zone) and No-Drone Zone (Airspace around airport at radius of 5 miles), there is a local regulation called KKOP (*Kawasan Keselamatan Operasi Penerbangan*). Refer to one of Indonesian Remote Pilot Certified and Activist, Arya Dega, the KKOP is a UAV restricted area that not included in NDZ or NFZ. *i.e.*: Presidential Palace, Government Building, some of Hospital, and Military Facility.

Japan

On 11 September 2015, an amendment to the Aeronautical Act was issued to introduce safety rules on unmanned aircraft (UA)/drones. The new regulations came into effect on 10 December 2015. According to the Civil Aviation Bureau in Japan, the term "UA/drone"s refers to any airplane, rotorcraft, glider or airship which cannot accommodate any person on board and can be remotely or automatically piloted (excluding those lighter than 200 grams, inclusive of the battery weight.)

Any person who intends to operate a UA/drones in the following airspace is required to obtain special permission from the Minister of Land, Infrastructure, Transport and Tourism.

- ✰ All airspace above or around airports, heliports and aerodromes. (airspace above the OLS, obstacle limitation surfaces.)
- ✰ All airspace 150 meters above ground/water surface.
- ✰ All airspace above Densely Inhabited Districts (DID), which are defined and published by the Ministry of Internal Affairs and Communications. (*i.e.* urban area and suburb, so only rural area is allowed)

When no permission is required from the minister, or obtained permission by extremely troublesome procedures, anyone should operate UA/drones in the daytime and within Visual Line of Sight (VLOS). Video telecontrol or automated flight is prohibited. A 30-meters operating distance have to be maintained between the UA/drones and persons or properties on the ground/water surface. UA/drones should not be operated over event sites where many people gather. UA/drones should not carry hazardous materials such as combustible or explosives (except one for power source). Objects should not be dropped off from UA/drones other than touching down the ground. These restrictions can be granted by special permission from the minister.

Furthermore, for security reasons, anyone cannot levitate any flying object including UA/drones, aeroballoons, gliders, flyboards and jet packs within about 300 meters from the National Diet Building, the official residents of speakers of the House of Representatives and the House of Councillors, the Prime Minister's Official Residence, the Kasumigasekis, the Supreme court, Tokyo Imperial Palace (including Akasaka Imperial Residences), the main offices of political parties, the diplomatic mission offices, the nuclear power plants and nuclear related facilities as well as buildings used for inter-governmental political forum such as Group of Seven temporarily. UA/drones will be radio-controlled, so one should follow the Japan Radio Act exactly. Flying UA/drones above some parks, coasts and ports including Tottori Sand Dunes is restricted by the by-laws in some prefectures of Japan. Like many other civilized countries, it is a kind of trespass to fly UA/drones indoors, near above the dwellings, or near above its surrounding soil without the permission of a person with legitimate rights.

Malaysia

Any person in charge of a small unmanned aircraft with a mass of 20 kilograms or less may fly the aircraft without approval from the Department of Civil Aviation Malaysia as long as "he is satisfied that the flight can safely be made"

Myanmar

In November 2017, Myanmar sentenced three journalists and their driver to two months' imprisonment for operating a drone near parliament while shooting a documentary. While Myanmar has no specific law on drones, individual authorities have tried to restrict the use of such equipment over their premises.

The Philippines

In 2015, the Civil Aviation Authority of the Philippines (CAAP) announced that owners or operators of unmanned aircraft vehicles (UAVs) who will operate the drones for commercial purposes must register with the CAAP, and secure a certification. Commercial uses include filming, aerial mapping, surveying and other revenue purposes. For operations specifically for model, sports, hobby and recreational activities, authorisation from the CAAP is not required. All operations of UAVs in Philippine airspace have to comply with the general guidelines stated in Philippine Civil Aviation Regulations Part 11.11.1.2 (General UAV Operations).

The **UAV must be flown**: 30 meters beyond a person who is not directly associated with the operation of the UAV; An altitude not exceeding 400 feet Above Ground Level (AGL); Outside 10 km radius from the Aerodrome Reference Point (ARP); and The UAV shall stay clear of populated area unless an approval from the CAAP has been granted.

Republic of Ireland

In 2012, the Irish Aviation Authority published a document setting out safety requirements for any unmanned aerial system. The IAA policy is that unmanned aerial systems may not be flown without the operator receiving a specific permission from the IAA. New regulations, including a registry of UAVs over 1 kg were introduced in December 2015.

Singapore

In Singapore, laws were passed in Parliament in May 2015 to allay concerns over safety, security and privacy surrounding unmanned aerial vehicles (UAVs). The Unmanned Aircraft (Public Safety and Security) Bill outlines regulations for the safe flying of drones and enforcement action against errant users. For instance, permits are required to fly drones above 7 kg, or within a 5 km radius of an aerodrome. Before conducting any outdoor activities, operators should ensure that the UA is flown within the permitted areas. The Civil Aviation Authority of Singapore (CAAS) website provides a map delineating prohibited areas, danger and restricted areas, areas within 5 km of an airport or an airbase and protected areas.

In general, permits are not required for recreational or research uses of UA in Singapore, as long as the operation of the UA is in line with CAAS' operating conditions. Situations where recreational or research uses of UA require a permit are when:

- The total mass of UA including payload exceeds 7 kg (Operator permit and Class 1 Activity Permit required)
- UA is flown higher than 200 feet above mean sea level (Class 2 Activity Permit required)
- Within restricted, danger, protected, prohibited areas and within 5 km of an aerodrome/airbase (Class 2 Activity Permit required)

Regardless of UA weight or location of UA operations, an operator permit and a Class 1 Activity Permit is required for operations that are non-recreational or non-research in nature. The permit application form can be found on the CAAS website. Depending on the nature of the flight or UA used, other possible permits that may be required from other agencies include: Singapore Police Force (SPF) for aerial photography and/or overflight of security-sensitive locations and Info-communications Media Development Authority (IMDA) of Singapore for use of radio frequencies and power limits other than in IMDA's guidelines for short range devices.

South Africa

In April 2014, the South African Civil Aviation Authority announced that it would clamp down on the illegal flying of UAVs in South African airspace. They also stated that as they had not authorised any such flights, existing ones were being done illegally.[36] A growth in the use of UAVs had prompted the SACAA to integrate them into South African airspace, but until regulations were in place, people operating them could be fined up to R50,000 and face up to 10 years prison.

Following discussions between the South African Civil Aviation Authority and key role players such as operators, manufacturers and other airspace users, a set of draft regulations were submitted to the Minister of Transport for review and approval. The regulations were accepted by Dipuo Peters, the South African transport minister, and put into effect 1 July 2015. The regulations are called the Eighth Amendment of the South African Civil Aviation Regulations, 2015.

The new regulations focus strongly on safety aspects with UAVs not being allowed to fly higher than 120m or closer than 50 meters to a person or group of people without prior approval from the SACAA.

Airspace restrictions also apply above and adjacent to:

- Nuclear plants
- Prisons
- Police stations
- Crime scenes
- Court of law
- National key points

Manned aircraft always have the right of way and the passing of UAVs close to such aircraft are strictly prohibited. Pilots are required to tune into the air traffic services in which they are flying and all flight activity are required to be recorded in a logbook. Public roads may also not be used for take-off and landing. Another safety aspect prohibits UAVs to be used in adverse weather conditions where visibility will be impaired. Separate approval is required where line of sight is unmanageable or for night flying.

Current regulations do not allow UAVs to be used for deliveries or for the transportation of goods. From a hobbyist point of view UAVs cannot tow other aircraft, perform aerobatic and aerial displays or be flown in a formation. UAV pilots are required to have a CAA approved and valid pilot license and a letter of approval which is valid for 12 months. Any incidents should be reported to the relevant authorities, especially if serious damage or injury was caused.

South Korea

Under the local law as of 2016, drones are banned from many places in the country, especially from the northern parts of Seoul, where key government offices are clustered. Areas around military installations and nuclear power plants are also no-fly zones

Thailand

All type of drones, except toy drones that are under 250 grams, have to be registered with the Civil Aviation Authority. The regulations cover nearly all forms of drone use from commercial and recreational to scientific. Drone users who failed to register their drones by 9 January 2018 could face up to five years in jail or a 100,000 baht (US$3100) fine.

United Kingdom

In August 2012, The U.K.'s Civil Aviation Authority (CAA) stated that it will require non-military drones larger than 20 kg to be able to automatically sense other aircraft and steer to avoid them. As of 2013, the CAA rules are that UAV aircraft less than 20 kilograms in weight must be in direct visual contact with the pilot, cannot fly within 150 meters of a congested area or within 50 meters of a person or vehicle, and cannot be used for commercial activity.

In July 2018, the CAA rules were updated to make it against the law to fly above 400 feet (120 m) and to make it against the law to fly your drone within 1 kilometre (0.62 mi) of an airport or airfield boundary. On February 20, 2019, The Department for Transport announced a new legislation to extend the 'no-fly' zone around airports, banning drones from flying within 5 kilometres (3.1 mi) of runways.

United States

The Federal Aviation Administration has adopted the name *unmanned aircraft* (UA) to describe aircraft systems without a flight crew on board. To operate a UA for non-recreational purposes in the United States, according to the FAA users must obtain a *Certificate of Authorization* to operate in national airspace. The FAA may permit the use of UAs for commercial or business purposes in response to individual exemption requests. In December 2015 the FAA announced that all UAVs weighing more than 250 grams flown for any purpose must be registered with the FAA. The FAA's Interim Rule can be accessed here. This regulation goes into effect on December 21, 2015 and requires that hobby type UAV's between 250 grams and 55 pounds need to be registered no later than February 19, 2016. The FAA's registration portal for drones can be accessed here.

Notable requirements of the new FAA UAV registration process include:

- ☆ Effective December 21, 2015, if the UAV has never been operated in U.S. airspace (*i.e.* its first flight outside), eligible owners must register their UAV's prior to flight. If the UAV previously operated in U.S. airspace, it must be registered by February 19, 2016.
- ☆ In order to use the registration portal, you must be 13 years of age or older. If the owner is less than 13 years old, then a parent or other responsible person must do the FAA registration.
- ☆ Each registrant will receive a certificate of aircraft registration and a registration number and all UAV's must be marked with the FAA issued registration number.
- ☆ The FAA registration requires a $5 fee, though the FAA is waiving the fee for the first 30 days of the new registration period - until January 20, 2016. The registration is good for 3 years, but can be renewed for an additional 3 years at the $5 rate.

The new FAA rule provides that a single registration applies to as many UAVs as an owner/operator owns or operates.[vague] Failure to register can result in civil penalties of up to $27,500, and criminal penalties which could include fines up to $250,000 and/or imprisonment for up to three years. To show problems with the FAA process, in August, 2015 an attorney was able to get FAA approval for a commercial drone that was actually a battery powered paper airplane toy. Its controllable range is 120 feet (37 meters) and maximum flight time is 10 minutes. It is too underpowered to carry a camera.[52] The December 2015 registration specifically excludes paper airplanes, Frisbees, and other flying objects that are not true UAVs. In addition to FAA certification, the regulation of usage of UA systems by government authorities in the United States for law enforcement purposes is determined at a state level.

Uruguay

In 2014, the government regulated UAVs. Machines under 25 kg do not require any operating licence, and have a maximum flight altitude of 120 meters within visual line of sight. Machines up to 260 kg must be registered, require an operator permit, and may have above 120 meters with permission from air traffic control. Machines above 260 kg are classified as aircraft, and require a full pilot license.

The regulatory guidelines for drone technology are highly variable and very generic. Major part of the guidelines set is for safety, surveillance and security and none of them related to agriculture. It is time, agricultural scientists begin to use and develop guidelines for large scale adoption of the technology.

Glossary

Agricultural drone is an unmanned aerial vehicle used to help optimize agriculture operations, increase crop production, and monitor crop growth. Sensors and digital imaging capabilities can give farmers a richer picture of their fields. Using an agriculture drone and gathering information from it may prove useful in improving crop yields and farm efficiency

Artificial Neural Networks (ANNs) or connectionist systems are computing systems vaguely inspired by the biological neural networks that constitute animal brains. Such systems "learn" to perform tasks by considering examples, generally without being programmed with task-specific rules.

Cohen's Kappa Coefficient is a statistic that is used to measure inter-rater reliability for qualitative items. It is generally thought to be a more robust measure than simple percent agreement calculation, as κ takes into account the possibility of the agreement occurring by chance.

Convolutional Neural Network (CNNs) is a class of deep neural networks, most commonly applied to analyzing visual imagery. They are also known as shift invariant or space invariant artificial neural networks, based on their shared-weights architecture and translation invariance characteristics.

Coronavirus disease 2019 is an infectious disease caused by severe acute respiratory syndrome coronavirus 2. It was first identified in December 2019 in Wuhan, Hubei, China, and has resulted in an ongoing pandemic.

Crop Water Stress Index (CWSI) – proposed by Jackson *et al.,* defined as the difference between air (Ta) and canopy temperature (Tc), normalized for the evaporative demand as determined by means of a lower limit (the case of a canopy transpiring at its potential rate) and an upper limit (a non-transpiring canopy), as shown in Equation.

CWSI = $(T_c - T_a) - (T_c - T_a)$ LL/$(T_c - T_a) - (T_c - T_a)$ UL

Where, **UL** and **LL** are the upper and lower limits, respectively.

Deep learning (DL) is part of a broader family of machine learning methods based on artificial neural networks with representation learning. Learning can be supervised, semi-supervised or unsupervised.

DGCI (Dark Green Colour Index) was used to assess the nitrogen concentration of the canopy, and it was calculated as:

DGCI = [H-60/60 + (1-S) + (1-B)]/3

Where H, S and B are the HSB of a digital image (converting R, G and B to H, S and B can be found in Karcher and Richardson's report (2003).

Drone (bee), a male bee, from an unfertilized egg and a fertile male ant

Drone - Unmanned aerial vehicle (UAV), an aircraft without a human aboard

Drone - Delivery type is an autonomous vehicle, often an unmanned aerial vehicle (UAV), used to transport packages, food or other goods

Eutrophication, distrophication or hypertrophication, is when a body of water becomes overly enriched with minerals and nutrients which induce excessive growth of algae. This process may result in oxygen depletion of the water body.

Fight velocity is a motion of the vehicle moving at a given direction in unit time. Its unit is meter/second

Flight height is a distance between the vehicle moving and above the ground level/ crop canopy. Its unit is meter

Fixed-wing aircraft is a flying machine, such as an airplane (or aeroplane; see spelling differences), which is capable of flight using wings that generate lift caused by the aircraft's forward airspeed and the shape of the wings. Fixed-wing aircraft are distinct from rotary-wing aircraft (in which the wings form a rotor mounted on a spinning shaft or "mast"), and ornithopters (in which the wings flap in a manner similar to that of a bird). The wings of a fixed-wing aircraft are not necessarily rigid; kites, hang gliders, variable-sweep wing aircraft and airplanes that use wing morphing are all examples of fixed-wing aircraft.

Geographic Information System is a conceptualized framework that provides the ability to capture and analyze spatial and geographic data. GIS applications are computer-based tools, that allow the user to create interactive queries, analyze spatial information output, edit datum presented within maps, and visually share the results of these operations.

Global Positioning System, originally NAVSTAR GPS, is a satellite-based radionavigation system owned by the United States government and operated by the United States Space Force.

GreenSeeker crop sensing system helps you effectively and precisely manage crop inputs on-the-go. With GreenSeeker, you can address field variability by applying the right amount of fertilizer, in the right place, at the right time.

Gyroscope is a device used for measuring or maintaining orientation and angular velocity. It is a spinning wheel or disc in which the axis of rotation is free to assume any orientation by itself. When rotating, the orientation of this axis is unaffected by tilting or rotation of the mounting, according to the conservation of angular momentum

Herbicides also commonly known as weedkillers, are substances used to control unwanted plants.

Hyperspectral imaging, like other spectral imaging, collects and processes information from across the electromagnetic spectrum. The goal of hyperspectral imaging is to obtain the spectrum for each pixel in the image of a scene, with the purpose of finding objects, identifying materials, or detecting processes.

Irrigation is the process of applying controlled amounts of water to plants at needed intervals. Irrigation helps to grow agricultural crops, maintain landscapes, and revegetate disturbed soils in dry areas and during periods of less than average rainfall.

Kettering Bug was an experimental unmanned aerial torpedo, a forerunner of present-day cruise missiles. It was capable of striking ground targets up to 121 kilometres (75 mi) from its launch point, while traveling at speeds of 80 kilometres per hour (50 mph).

MATLAB is a multi-paradigm numerical computing environment and proprietary programming language developed by MathWorks. MATLAB allows matrix manipulations, plotting of functions and data, implementation of algorithms, creation of user interfaces, and interfacing with programs written in other languages.

Micro air vehicle (MAV), or micro aerial vehicle, is a class of miniature UAVs that has a size restriction and may be autonomous. Modern craft can be as small as 5 centimeters. Development is driven by commercial, research, government, and military purposes; with insect-sized aircraft reportedly expected in the future. The small craft allows remote observation of hazardous environments inaccessible to ground vehicles. MAVs have been built for hobby purposes, such as aerial robotics contests and aerial photography

Miniature UAV or small UAV (SUAV) is an unmanned aerial vehicle small enough to be man-portable.

Miniature UAVs range from micro air vehicles (MAVs) that can be carried by an infantryman, to man-portable UAVs that can be carried and launched like an infantry man-portable air-defense system. The term is usually applied to those used for military purposes.

Multirotor or multicopter is a rotorcraft with more than two lift-generating rotors. An advantage of multirotor aircraft is the simpler rotor mechanics required for flight control. Unlike single- and double-rotor helicopters which use complex variable pitch rotors whose pitch varies as the blade rotates for flight stability and control, multirotors often use fixed-pitch blades; control of vehicle motion is achieved by varying the relative speed of each rotor to change the thrust and torque produced by each.

Multispectral image is one that captures image data within specific wavelength ranges across the electromagnetic spectrum. The wavelengths may be separated by filters or detected via the use of instruments that are sensitive to particular wavelengths, including light from frequencies beyond the visible light range

NGRDI - Normalized Green Red Difference Index

NGRDI = (Green DN) Red DN)/(Green DN + Red DN)

Object-Based Image Analysis (OBIA) employs two main processes, segmentation and classification. Traditional image segmentation is on a per-pixel basis. However, OBIA groups pixels into homogeneous objects. These objects can have different shapes and scale. Objects also have statistics associated with them which can be used to classify objects.

Orthophoto, orthophotograph or orthoimage is an aerial photograph or image geometrically corrected such that the scale is uniform: the photo has the same lack of distortion as a map.

Parafoil is a nonrigid airfoil with an aerodynamic cell structure which is inflated by the wind. Ram-air inflation forces the parafoil into a classic wing cross-section. Parafoils are most commonly constructed out of ripstop nylon.

Passenger drone (also known as a drone taxi, flying taxi, or pilotless helicopter) is a type of personal air vehicle (PAV). The first passenger drone was introduced at the Consumer Electronics Show (CES) 2016 by Chinese entrepreneurs and is called the Ehang 184.

Performance assessment

To assess the performance of the detection method quantitatively, three indices were evaluated, *i.e.*, first the **M index** which is the ratio of weed pixels to the total number of pixels of the image.

$$\text{M (per cent)} = \frac{\text{Image Number of Weed Pixels}}{\text{Total Image Number of Pixels}} \times 100 \text{ per cent}$$

Second, the **M/MGT index**, which is the ratio of the number of pixels that were detected as weed into the image in grayscale to the pixels that were selected as weed into the ground truth image. Hence, the ratio M/MGT indicates the percentage of pixels found to be weed.

$$M/MGT(\text{per cent}) = \frac{\text{Image Number of Weed Pixels}}{\text{Ground Truth Number of Weeds Pixels}} \times 100 \text{ per cent}$$

Third, the MP index, the ratio of the number of matching weed pixels between the ground truth image and the analyzed image to the total number of weed pixels of the image in grayscale was calculated. This index reveals the percentage of correct weed pixel detection. Additionally, to compute the **MP index**, a pixel matching algorithm was used. The algorithm compares the values of pixels at the same co-ordinate for the ground-truth and the images obtained from NN classification system.

$$MP(\text{per cent}) = \frac{\text{Number of matching Pixels}}{\text{Ground Truth Number of Weeds Pixels}} \times 100 \text{ per cent}$$

WCE - weed control efficiency

Evaluation of droplet deposition in aerial spraying operation indexes could be calculated

a. **Droplet deposition density** (D) = n/A

n is the number of droplets deposited on the collection material

A is the area of droplet collection material.

b. **Droplet deposition coverage** (C) = S/A

S is the area of droplet deposition particle

A is the area of droplet collection material

c. **Arithmetic mean of droplet size** (D_0) = Σ Di Ni/Σ Ni

Di is the droplet diameter of a certain size interval

Ni is the number of droplets at a certain interval

d. **Variation coefficient CV** = SD/X

$$\text{Standard Deviation (SD)} = \sqrt{\frac{\sum_{i=1}^{n}(x_i - \mu)^2}{n-1}}$$

Xi is the droplet deposition particle size per unit area of each sample card

μ is the average droplet size per unit area of each sample card

SD is the standard deviation

n is the number of droplets per sample card.

The smaller the variation coefficient *CV*, the more uniform the droplet distribution in the target spraying area, and the better the pesticide effect plays. X

The population decline rate and control efficacy were calculated by the following formulas:

Population Decline Rate (R) = (Nb - Na)/Nb 9 100 per cent,

Control Efficacy (E) = (Rt - Rc)/(1 - Rc) 9 100 per cent,

where **Nb** represents the number of alive FAW before spraying, **Na** represents the number of alive FAW after spraying, **Rt** represents the population decline rate in the treatment and **Rc** represents the population decline rate in the control.

QGIS is a free and open-source cross-platform desktop geographic information system application that supports viewing, editing, and analysis of geospatial data.

Quadcopter or quadrotor is a type of helicopter with four rotors. Quadcopters generally have two rotors spinning clockwise (CW) and two counterclockwise (CCW). Flight control is provided by independent variation of the speed and hence lift and torque of each rotor. Pitch and roll are controlled by varying the net centre of thrust, with yay controlled by varying the net torque.[

Remote sensing is the acquisition of information about an object or phenomenon without making physical contact with the object and thus in contrast to on-site observation, especially the Earth.

RGB-based Vegetation Indices

Excess Greenness Index (E x G)2 x G – R - B

Normalized Difference Index NDI   G - R/G + R

NIRL: Reflectance values in the near-infrared band

G: Reflectance values from the green band

Spatial Resolution in Remote Sensing describes how much detail in a photographic image is visible to the human eye. The ability to "resolve," or separate, small details is one way of describing what we call spatial resolution. Spatial Resolution of images acquired by satellite sensor systems is usually expressed in meters

Spray Drift Herbicide/Pesticide spray drift is the aerial movement, and unintentional deposit, of pesticide outside the target area.

Spray volume is the amount spray fluid needed to spray an unit area of crop canopy. Its unit is liters/unit area (lit/ha.)

Unmanned combat aerial vehicle (UCAV), also known as a **combat drone**, is an unmanned aerial vehicle (UAV) that usually carries aircraft ordnance such as missiles, ATGMs, and/or bombs and is used for drone strikes. These drones are usually under real-time human control, with varying levels of autonomy. Aircraft of this type have no onboard human pilot. As the operator runs the vehicle from a remote terminal, equipment necessary for a human pilot is

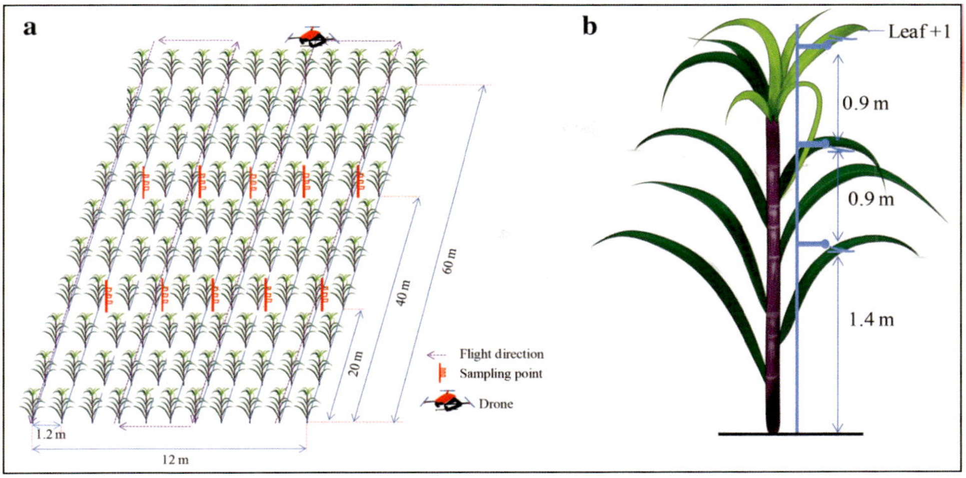

Drones Direction of Flights and Sampling Points. (Pg. 113)

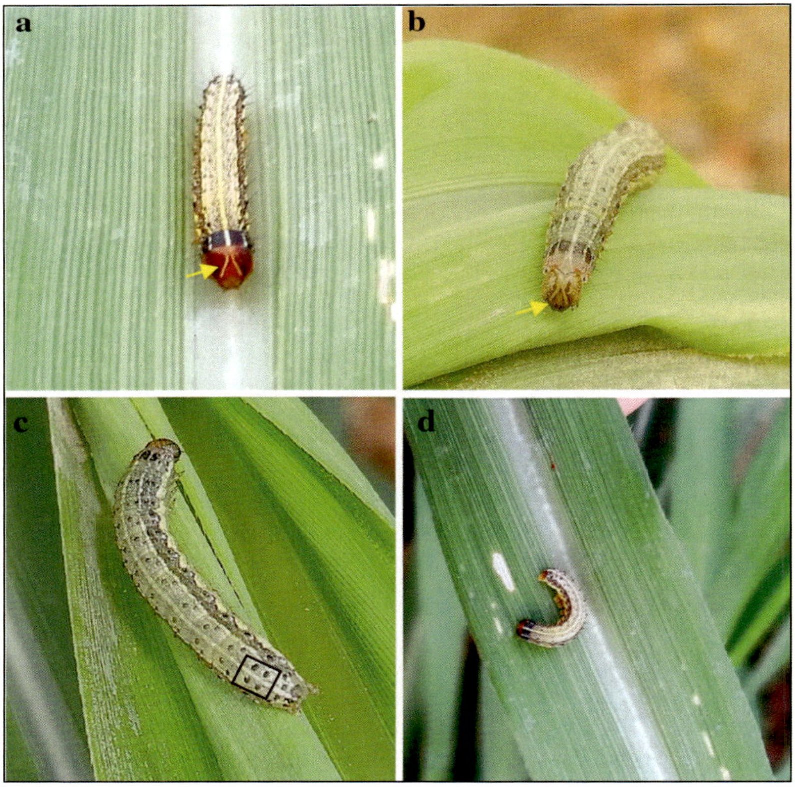

Morphological Characteristics of Fall Armyworm. (a, b) Show the 'Y' shape on the head; (c) Shows the four black flecks and d shows the thanatosis. (Pg 114)

Damage Symptoms of Sugarcane by Fall Armyworm. (a) Shows the leaf epidermis; (b) Shows the irregular holes and incisions; (c) Shows the frass; (d) Shows the damaged whorl and ? indicates the positions ofdamage symptoms. (Pg 115)

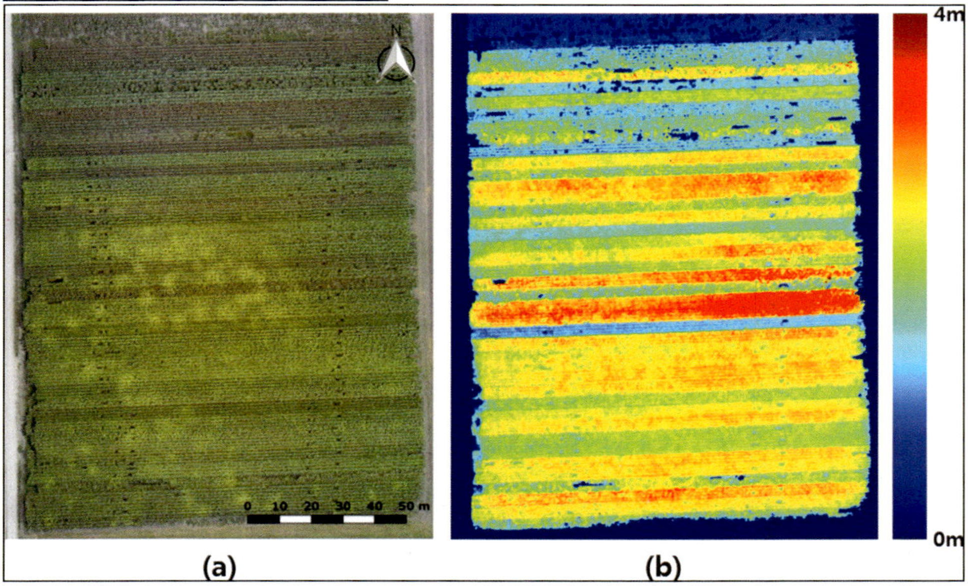

Images from the Study Area of (a) Ortho-mosaic and (b) Crop height model (CHM) of SORGHUM generated from UAS data acquired on July 24, 2016 (color). (For interpretation of the references to colour in this figure legend, the reader is referred to the web version of this article). (Pg 131)

The Third Analysis is to Identify the Water Flow in the Paddy Plot Area. Because there's an uneven ground in the paddy plot, it will lead the inconsistent of water level condition. Since the water level should be 5 cm levelling, the plants at the area will be affected because of the flooding and their development will be stunted, consequence the declining yields production during the harvesting time. This analysis is important to make sure the water flow is in a good direction. (*Source*: Norasma *et al.,* 2018). (Pg 132)

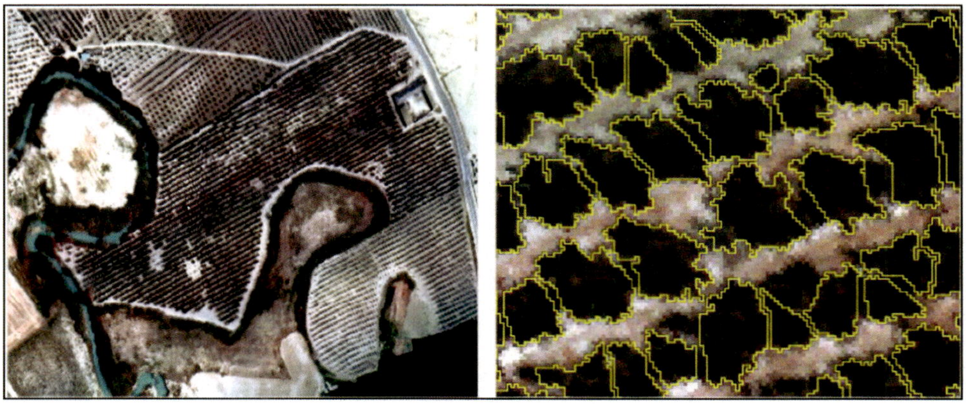

Multispectral Scene (a) obtained with the multispectral camera on board the UAV platform at 20-cm resolution, showing the Castro del Rio orchard study site (Córdoba province). (b) Automatic object-based crown detection applied to the multispectral imagery to identify pure olive crowns. Yellow square (a) shown in detail in (b). (For interpretation of the references to color in this figure legend, the reader is referred to the web version of this article.) (Pg 135)

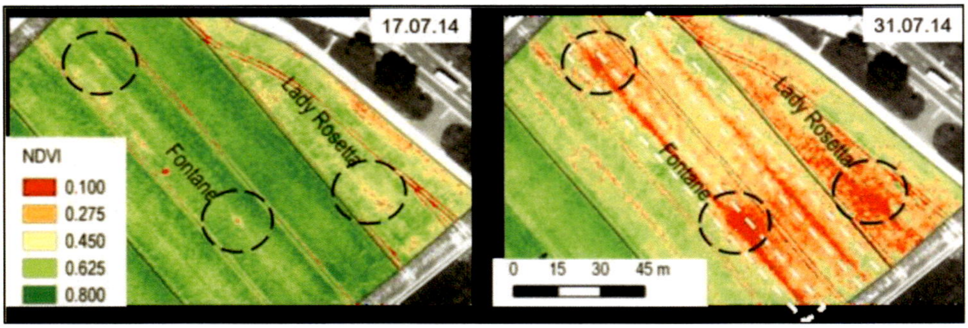

Detail View of Potato Field NDVI Map Highlighting the Three Original Sites of Potato Blight Infestation (Dotted circles) and the Subsequent Spreading of the Infestation (Pg 136)

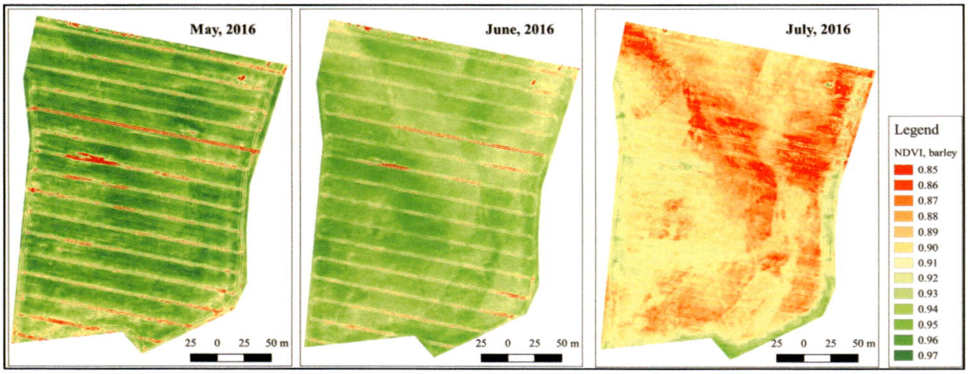

Barley: NDVI Maps, May-July, 2016. (Pg 137)

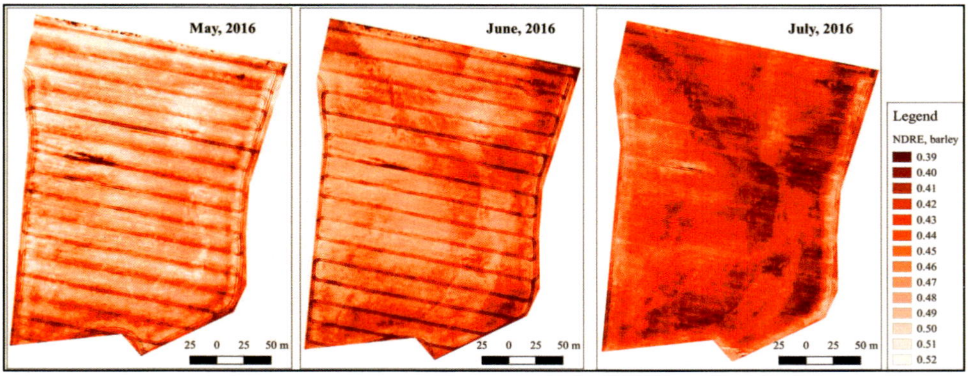

Barley: NDRE Maps, May-July, 2016. (Pg 138)

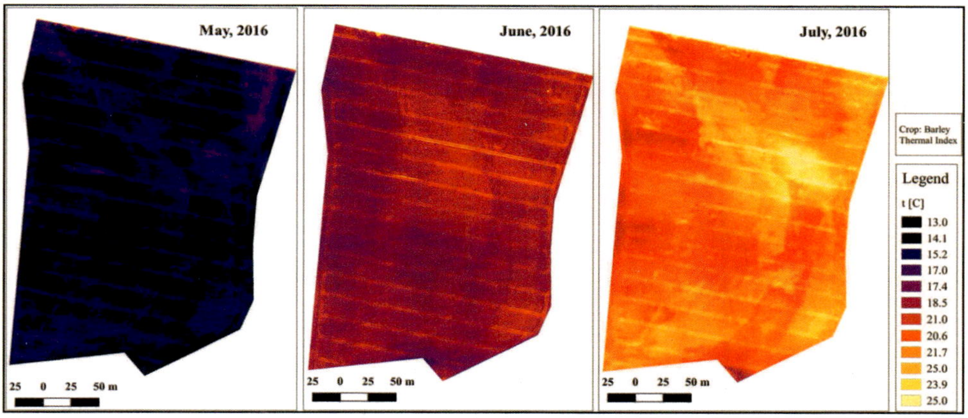

Barley: Thermal Maps, May-July, 2016. (Pg 138)

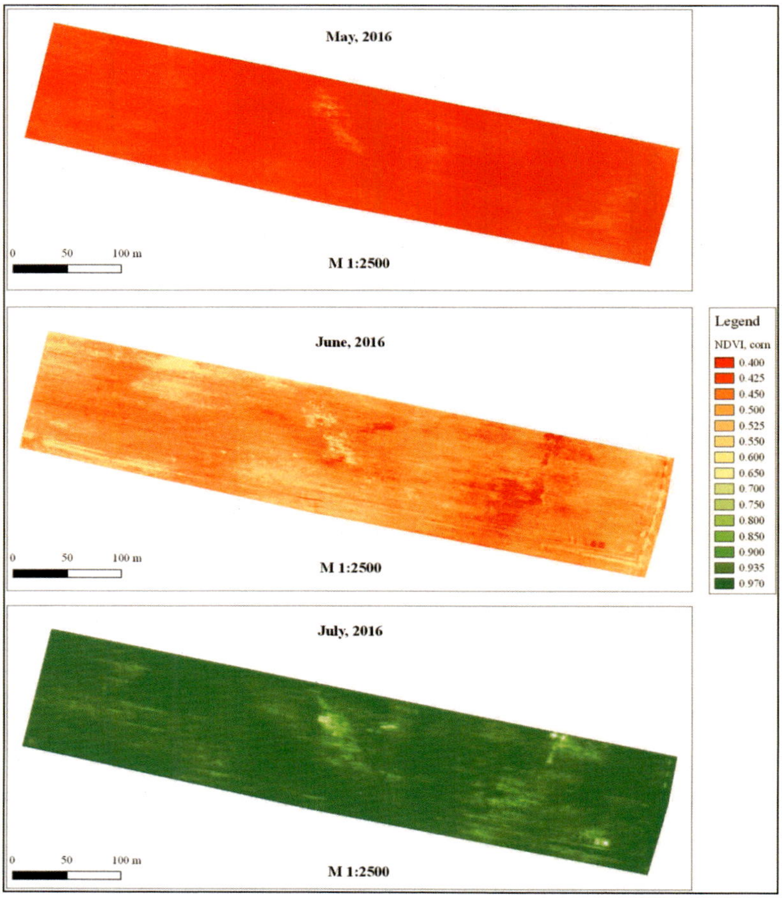

Figure 9.8. Corn: NDVI Maps, May-July, 2016. (Pg 139)

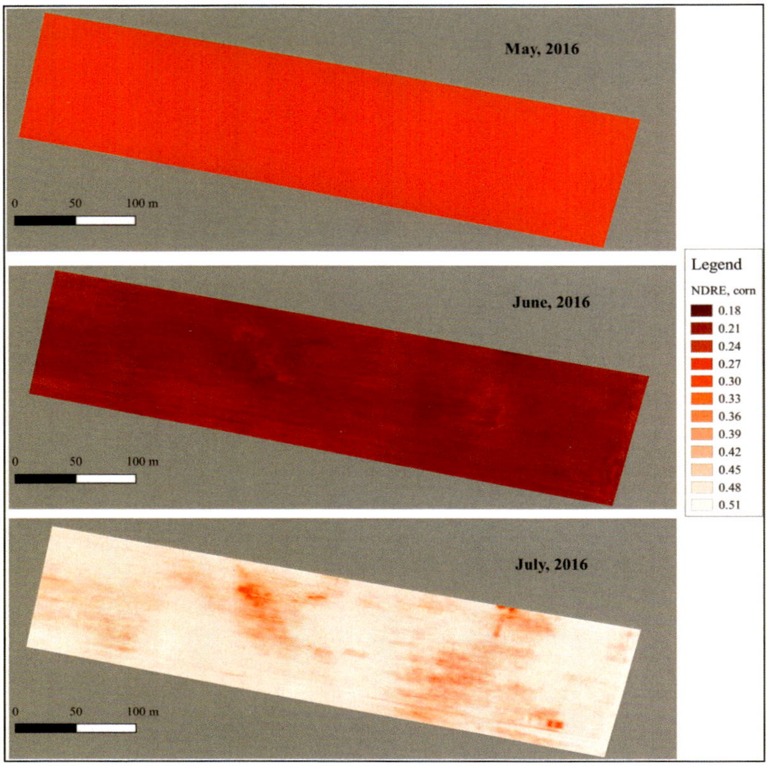

Corn: NDRE Maps, May-July, 2016. (Pg 140)

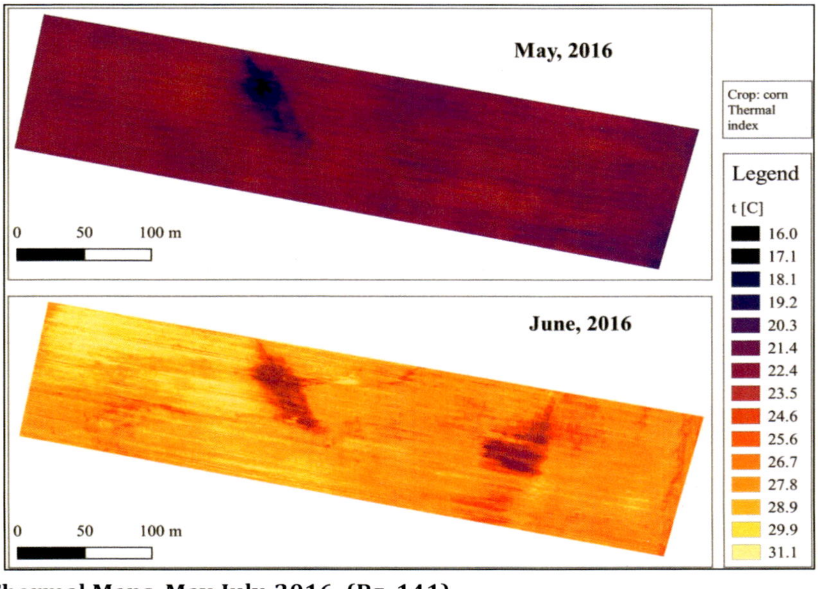

Corn: Thermal Maps, May-July, 2016. (Pg. 141)

Location of the Study site within New Zealand and an Image of the Coaxial Quad-copter UAV used during Data Collection. The main panel shows an image of the multispectral data collected with treated trees evident and outlined in blue (image date = 03-03-2016). (For interpretation of the references to colour in this figure legend, the reader is referred to the web version of this article) (*Source*: Dash *et al.*, 2017). (Pg 142)

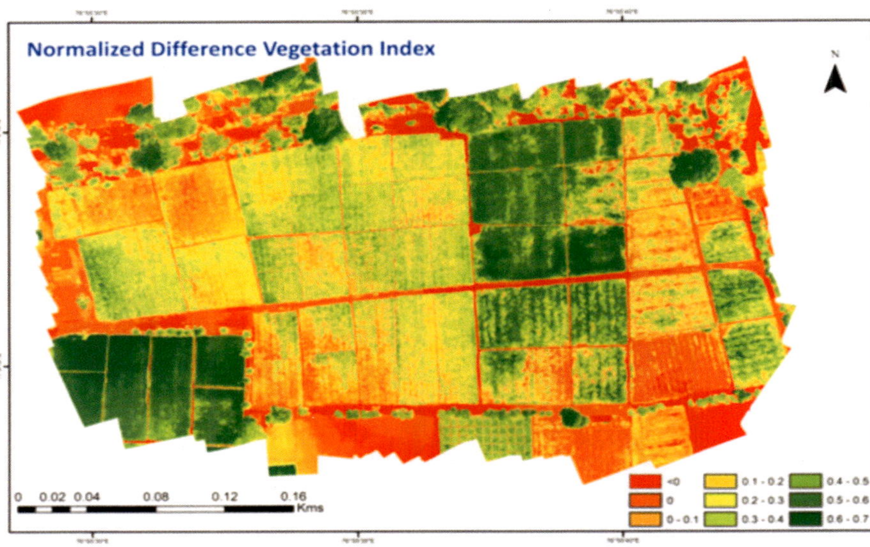

NDVI map of Wetland farm, TNAU, Coimbatore (Pg. 145)

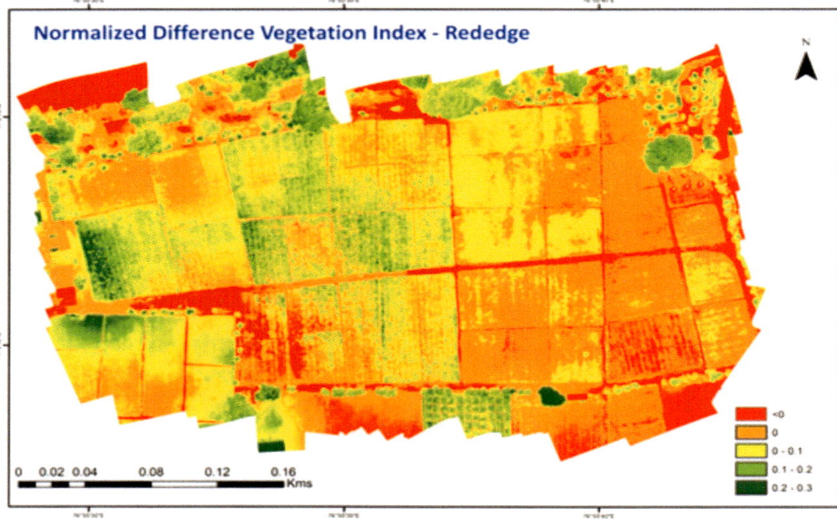

NDVI-Rededge map of Wetland farm, TNAU, Coimbatore (Pg. 146)

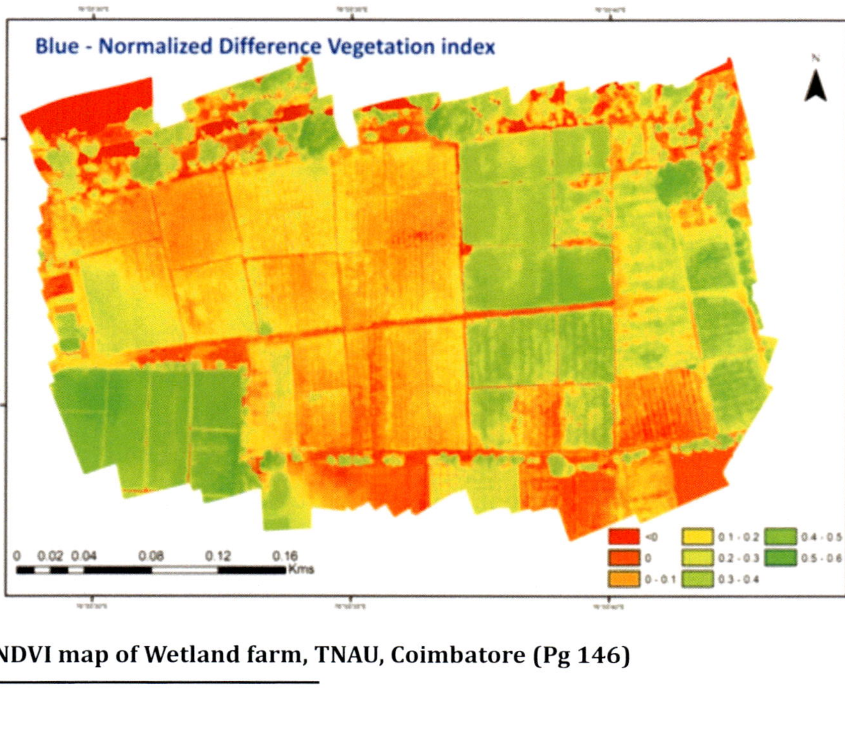

Blue-NDVI map of Wetland farm, TNAU, Coimbatore (Pg 146)

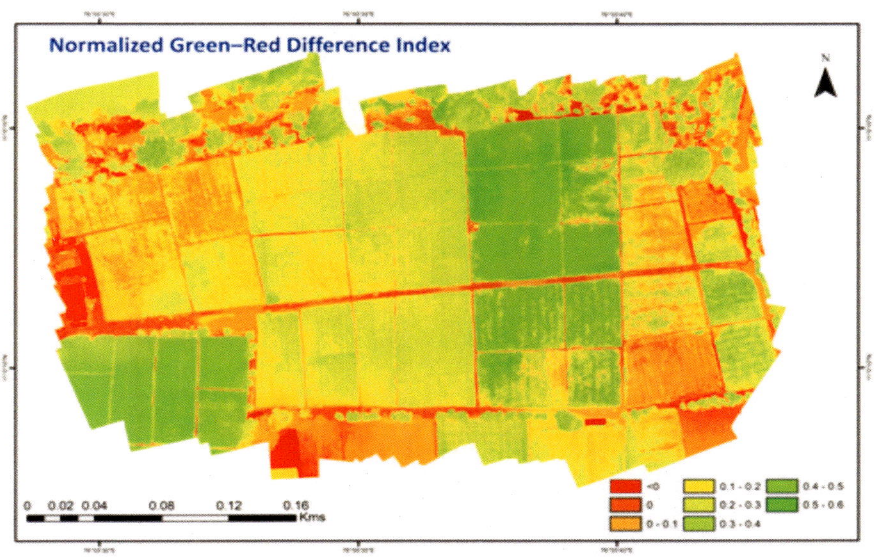

Normalized Green-Red Difference Index map of Wetland farm, TNAU, Coimbatore (Pg 147)

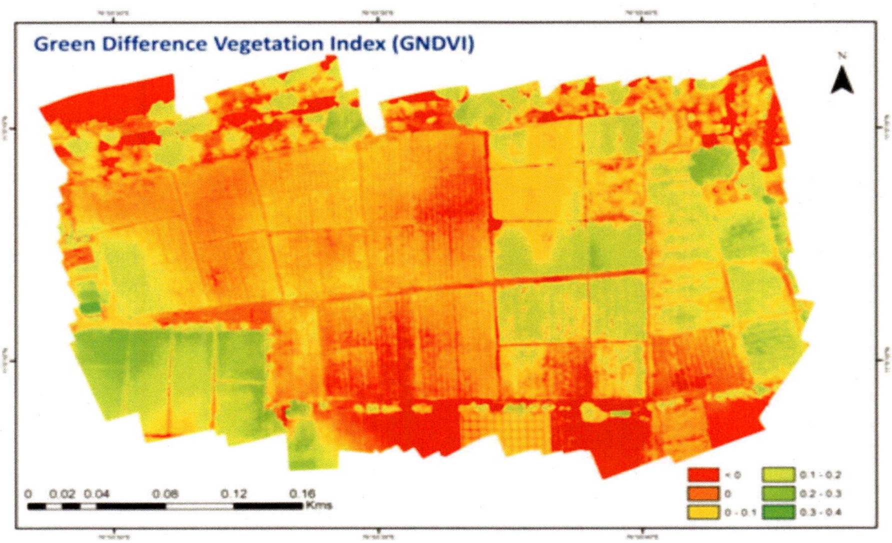

Green-NDVI map of Wetland farm, TNAU, Coimbatore (Pg. 148)

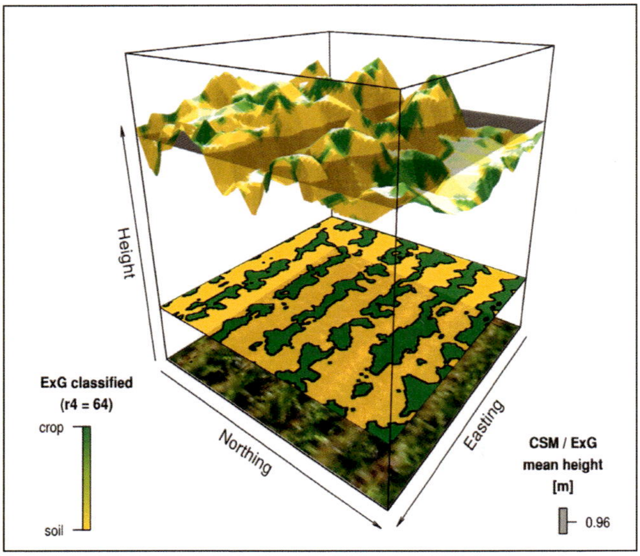

Mean Crop Height Computation using the Example of a 4 x 4 m Sub-sample of Plot 413 with a Sowing Density of 11 Seeds_m^{-2} and Nitrogen Application of 50 kg_N_ ha^{-1}. The lower part of the figure shows a stack of the RGB orthoimage and the ExG layer classification based on threshold r4 at a ground resolution of 0.04 m and at crop growth stage Z39. The upper part shows the corresponding CSM layer height information as a 3D representation, colored by the ExG-classification. Mean crop height was calculated by the crop-classified CSM layer heights only and is displayed as a semi-transparent plane (*Source:* Geipel *et al.*, 2014). (Pg 154)

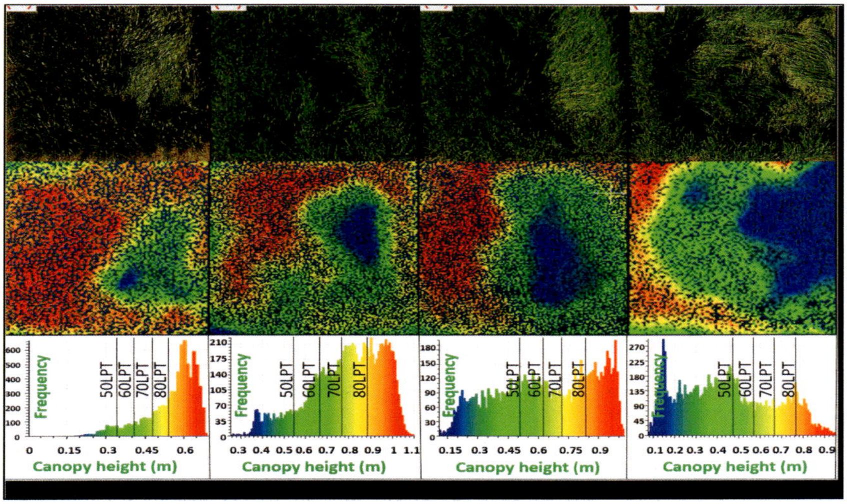

RGB Images of Barley Plots showing different Types (a–d) of Lodging (Top) and Corresponding Lodging Severity Derived from the CHM (Middle), as well as Canopy Height Distributions with Visualization of different Lodging Percentage Thresholds (LPTs) (Bottom) (Source: Wilke *et al.*, 2019). (Pg 157)

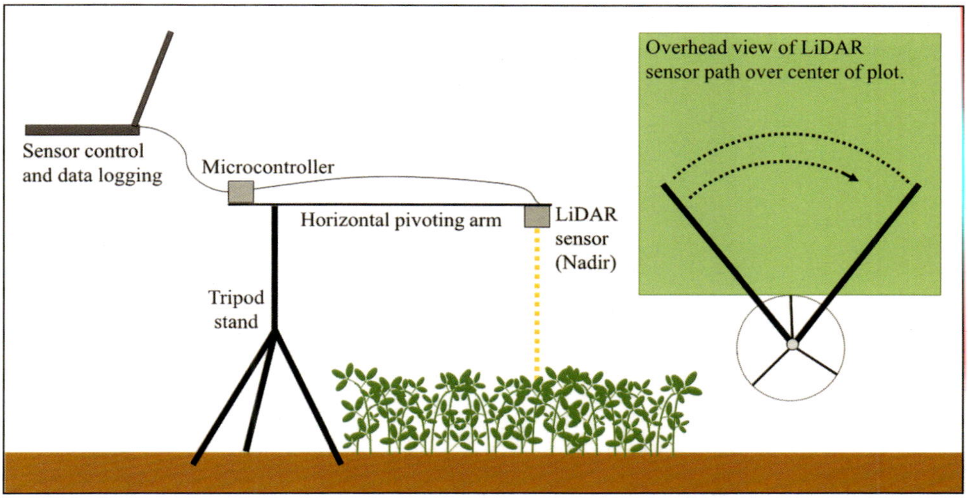

Illustration of LiDAR Sensor Deployment to Estimate Canopy Height in Alfalfa Plots (Source: Noland *et al.,* 2018). (Pg 161)

Elephant-Human Conflict: A Cause for Concern. (Pg. 169)

(a) Glossy ibis (*Plegadis falcinellus*) individual. (b) Front view of the Glossy ibis colony, densely covered by the macrophyte Typha, which makes it extremely difficult ground census of the colony. (Pg. 174)

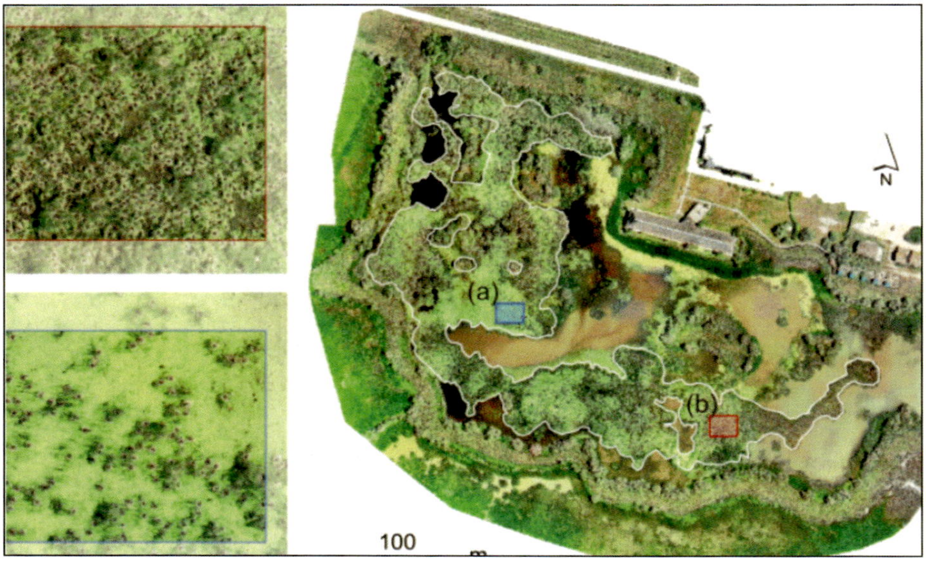

Orthomosaic Image of High-resolution of the Glossy Ibis Colony in "Lucio de la FAO". Image was mosaicked with Pix4D mapper Pro (Pix4D, Switzerland) after acquiring 448 photos through automatic flight of Phantom 4 Pro quadcopter (DJI Innovations, Shenzhen, China) on 14 May 2018. Squares show validation areas to test the accuracy of the automatic nest counting method to the visually nest count by an expert. Blue squares show densely macrophytes area (Typha spp) whereas red squares showfree areas of emergent vegetation covered by Lemna aquatic plant. White line delineates the Glossy ibis nesting area. (Pg. 175)

not needed, resulting in a lower weight and a smaller size than a manned aircraft. Many countries have operational domestic UCAVs and many more have imported armed drones or have development programs underway.

Underwater glider is a type of autonomous underwater vehicle (AUV) that employs variable-buoyancy propulsion instead of traditional propellers or thrusters. It employs variable buoyancy in a similar way to a profiling float, but unlike a float, which can move only up and down, an underwater glider is fitted with hydrofoils (underwater wings) that allow it to glide forward while descending through the water.

Vegetation Index is a spectral transformation of two or more bands designed to enhance the contribution of vegetation properties and allow reliable spatial and temporal inter-comparisons of terrestrial photosynthetic activity and canopy structural variations.

Vegetation Indices derived from multispectral information

Ratio Vegetation Index (RVI) NIR/R

Normalized Difference Vegetation Index (NDVI) NIR – R/NIR + R

Normalized Difference Red Edge Index (NDRE) NIR – RE/NIR + RE

Green NDVI (GNDVI) NIR – G/NIR + G

Water Deficit Index (WDI) is calculated as defined by Moran *et al.* (1994) as:

$$WDI = 1 - \lambda Eact \; / \; \lambda Epot = (TS - TA)min - (TS - TA)mes \; / \; (TS - TA)min - (TS - TA)max$$

λEact is the actual latent heat flux density (W m^{-2})

λEpot is the potential latent heat flux density (W m^{-2})

TS is the composite LST land surface temperature (°C),

TA is the air temperature (°C)

Subscripts 'min', 'max' and 'mes' refer to minimum, maximum and measured.

References

Acorsi, MG,Fabiani das DoresAbati Miranda, Maurício Martello, Danrley Antonio Smaniotto and Laercio Ricardo Sartor. 2019. "Estimating Biomass of Black Oat Using UAV-Based RGB Imaging". *Agronomy* 9: 344.

Afan, I, M Máñez and Ricardo Díaz-Delgado. 2018. "Drone Monitoring of Breeding Waterbird Populations: The Case of the Glossy Ibis". *Drones* 2: 42. doi:10.3390/drones2040042

Agarwala, M, S Kumar, A Treves and L Naughton-Treves. 2010."Paying for wolves in Solapur, India and Wisconsin, USA: Comparing compensation rules and practice to understand the goals and politics of wolf conservation". *Biological Conservation Journal*143: 2945-2955.

Agüera, F, F Carvajal and M Pérez. 2011. "Measuring Sunflower Nitrogen Status from An Unmanned Aerial Vehicle-Based System and Anon The Ground Device". *International Archives of the Photogrammetry, Remote Sensing Spatial Information Sciences 3822*: 33-37.

Agüera, F, F Carvajal and M Pérez. 2011. "Measuring sunflower nitrogen status from an unmanned aerial vehicle-based system and an on the ground device". *The International Archives of the Photogrammetry, Remote Sensing and Spatial Information Sciences,* 38: 33-37.

Akira Hama, Kei Tanaka, Atsushi Mochizuki, Hiroyuki Arai, Toshiyuki Hirata, RyuyaYawata, Eunice Nduati, YasuoTsuruoka and Akihiko Kondoh.2017. "Estimation of Rice Yield Based on the Integration of UAV Remote Sensing and Solar Radiation Data". Japan Geoscience Union.

Analysis". *Nature Methods* 9(7): 671-675.

Anderson, K and KJ Gaston. 2013. "Lightweight unmanned aerial vehicles will revolutionize spatial ecology". *Frontiers in Ecology and the Environment* 11: 138-146. https://doi.org/10.1890/120150

Aning, J, H Wenjiang, Z Chunjiang. 2007. "Effects of variable nitrogen application based on characteristics of canopy light reflectance in wheat". *SciAgric Sin.* 40(9): 1907-13.

Araus, JL and JE Cairns. 2014. "Field high-throughput phenotyping: the new crop breeding frontier". *Trends Plant Science* 19: 52-61. http://dx.doi.org/10.1016/j.tplants.2013.09.008.

Baatz, M, U Benz, S Dehghani, M Heynen, A Holtje, P Hofmann, I Lingenfelder, M Mimler, M Sohlbach and M Weber. 2011. "eCognition Professional User Guide 4; A Definiens Imaging: Munich", Germany, 2004.

Bae, Y and YM Koo. 2013. "Flight attitudes and spray patterns of a roll-balanced agricultural unmanned helicopter". *Applied Engineering in Agriculture* 29: 675-682

Baena, S, J Moat, O Whaley and DS Boyd. 2017. "Identifying species from the air: UAVs and the very high resolution challenge for plant conservation". *PLoS ONE,*12, e0188714.

Bah, MD, A Hafiane and R Canals. 2018. "Deep Learning with Unsupervised Data Labeling for Weed Detection in Line Crops in UAV Images". *Remote Sensing* 10: 1690.

Bah, MD, E Dericquebourg, A Hafiane and R Canals. 2018. "Deep Learning Based Classification System for Identifying Weeds Using High-Resolution UAV Imagery. In Proceedings of the Science and Information Conference", London, UK, 10–12 July 2018; Springer: Cham, Switzerland, 2018; 176-187.

Ballester, C, J Horn buckle, J Brinkhoff, J Smith and W Quayle. 2017. "Assessment of in-season cotton nitrogen status and lint yield prediction from unmanned aerial system imagery". *Remote Sensing 9*: 1149.

Balota, M and J Oakes. 2017. UAV Remote Sensing for Phenotyping Drought tolerance In Peanuts". In *Autonomous Air and Ground Sensing Systems for Agricultural Optimization and Phenotyping II*; Society of Photo-Optical Instrumentation Engineers (SPIE) Conference Series; International Society for Optics and Photonics: Washington, DC, USA, 10218: 102180C.

Bannari, A, D Morin, F Bonn and A Huete. 1995. "A review of vegetation indices". *Remote Sensing Reviews* 13: 95-120.

Barrero, O and SA Perdomo. 2018. "RGB and multispectral UAV image fusion for Gramineae weed detection in rice fields". *Precision Agriculture* 19(5): 809-22. doi:10.1007/s11119-017-9558-x.

Barua, M, SA Bhagwat and S Jadhav. 2013. "The hidden dimensions of human-wildlife conflict: Health impacts, opportunity and transaction costs". *Biological Conservation Journal*157: 309-316.

Basso, B, JT Ritchie, D Cammarano. 2011. "A strategic and tactical management approach to select optimal N fertilizer rates for wheat in a spatially variable field". *European Journal of Agron*omy 35(4): 215-22.

Bausch, WC, AD Halvorson andJ Cipra. 2008. "Quickbird satellite and ground-based multispectral data correlations with agronomic parameters of irrigated maize grown in small plots". *Biosystems Engineering*101: 306-315.

Becker-Reshef, I, E Vermote, M Lindeman. 2010. "A generalized regression based model for forecasting winter wheat yields in Kansas and Ukraine using MODIS data". *Remote Sensing Environment* 114: 1312-1323. doi: 10. 1016/j rse.2010.01.010

Bellvert, J, PJ Zarco-Tejada, J Girona and E Fereres. 2014. "Mapping Crop Water Stress Index In A 'Pinot-noir' Vineyard: Comparing Ground Measurements with Thermal Remote Sensing Imagery From An Unmanned Aerial Vehicle". *Precision Agriculture* 15: 361-376.

Benincasa, P, S Antognelli, L Brunetti, CA Fabbri, A Natale, V Sartoretti, G Modeo, M Guiducci, F Tei and M Vizzari. 2018. "Reliability of NDVID erived by High Resolution Satellite and UAV Compared to In-field Methods for The Evaluation of Early Crop N Status and Grain Yield In Wheat". *Experimental Agriculture* 54: 604-622.

Benitez Ramirez, M. 2010. "Monitoring nitrogen levels in the cotton canopy using real-time active-illumination spectral sensing". University of Tennessee-Knoxville.

Berger-Neto, A, DDS Jaccoud-Filho, CR Wutzki, HE Tullio, MLC Pierre, F Manfron. 2017. "Effect of spray droplet size, spray volume and fungicide on the control of white mold in soybeans". *Crop Protection* 92: 190-197.

Berni, J, PJ Zarco-Tejada, L Suarez and E Fereres. 2009. "Thermal and Narrowband Multispectral Remote Sensing for Vegetation Monitoring from an Unmanned Aerial Vehicle. IEEE Trans". *Geoscience and Remote Sensing* 47: 722-738.

Berni, JAJ, PJ Zarco-Tejada, G Sepulcre-Cantó, E Fereres and F Villalobos. 2009. "Mapping canopy conductance and CWSI in olive orchards using high resolution thermal remote sensing imagery". *Remote Sensing of Environment* 113: 2380-2388.

Bhandari, S, A Raheja, MR Chaichi, RL Green, D Do, FH Pham, M Ansari, JG Wolf, TM Sherman and A Espinas. 2018. "Effectiveness of UAV-Based Remote Sensing Techniques in Determining Lettuce Nitrogen and Water Stresses". In Proceedings of the 14th International Conference in Precision Agriculture, Montreal, QC, Canada, 24-27 June 2018.

Bhandari, S, A Raheja, MR Chaichi, RL Green, D Do, FH Pham, M Ansari, JG Wolf, TM Sherman and A Espinas. 2018. "Lessons Learned from UAV-Based Remote Sensing for Precision Agriculture". In Proceedings of the 2018 International Conference on Unmanned Aircraft Systems (ICUAS), Dallas, TX, USA, 12–15 June 2018; 458–467.

Bierman, PM, CJ Rosen and RT Venterea. 2012. "Survey of nitrogen fertilizer use on corn in Minnesota". *Agricultural systems* 109(6): 43-52.

Biswas, PP and PD Sharma. 2008. "A new approach for estimating fertilizer response ratio-the Indian". *Indian Journal of Fertilisers* 4 (7): 59-62

Boegh, E, H Soegaard, N Broge. 2002. "Airborne multispectral data for quantifying leaf area index, nitrogen concentration, and photosynthetic efficiency in agriculture". *Remote Sensing of Environment* 81(2-3): 179-93.

Bonnin, N, A Van Andel, J Kerby, A Piel, L Pintea and S Wich. 2018. "Assessment of Chimpanzee Nest Detectability in Drone-Acquired Images". *Drones*2: 17.

Bonotis, PA, DC Tsouros, PN Smyrlis, AT Tzallas, N Giannakeas, G Evripidis and MG Tsipouras. 2019. "Automated Assesment of Pain Intensity based on EEG Signal Analysis". In Proceedings of the IEEE 19thInternational Conference on BioInformatics and BioEngineering, Athens, Greece, 28–30 October 2019.

Bregoli, F, A Crosato, P Paron and M McClain. 2019."Humans reshape wetlands: Unveiling the last 100 years of morphological changes of the Mara Wetland, Tanzania". *Science of the Total Environment*691: 896-907.

Burke, M and DB Lobell. 2017."Satellite-based assessment of yield variation and its determinants in smallholder African systems". *Proceedings of the National Academy of Sciences of the United States of America* 114: 2189-2194.

Busemeyer, L, D Mentrup, K Möller. 2013. "Breed Vision-a multi-sensor platform for non-destructive field-based phenotyping in plant breeding". *Sensors* 13(3): 2830-47.

Buters, T, David Belton and Adam Cross. 2019. "Seed and Seedling Detection Using Unmanned Aerial Vehicles and Automated Image Classification in the Monitoring of Ecological". *Drones* 3(3): 53.

Buters, TM, PW Bateman, T Robinson, D Belton, KW Dixon and AT Cross.2019. "Methodological ambiguity and inconsistency constrain unmanned aerial vehicles as a silver bullet for monitoring ecological restoration". *Remote Sensing* 11.

Cai, G, J Dias and L Seneviratne. 2014. "A survey of small-scale unmanned aerial vehicles: Recent advances and future development trends". *Unmanned Systems* 2: 175-199.

Calderón, R, J Navas-Cortés, C Lucena, and P Zarco-Tejada. 2013. "High-resolution Airborne Hyperspectral and Thermal Imagery for Early Detection of

Verticillium Wilt of Olive Using Fluorescence, Temperature and Narrow-band Spectral Indices". *Remote Sensing Environment* 139: 231-245.

Candiago, S, F Remondino, M De Giglio, M Dubbini and M Gattelli. 2015. "Evaluating Multispectral Images and Vegetation Indices for Precision Farming Applications from UAV Images". *Remote Sensing* 7: 4026-4047.

Cao, LP. 2015. "The Research Progress on Machine Recognition of Plant Diseases and Insect Pests". *Chinese Agricultural Science Bulletin* 31(20): 244-249.

Capolupo, A, L Kooistra, C Berendonk, L Boccia and J Suomalainen. 2015. "Estimating Plant Traits of Grasslands From UAV-Acquired Hyperspectral Images: A Comparison of Statistical Approaches". *ISPRS International Journal of Geo-Information* 4: 2792–2820.

Carlson, TN and DA Ripley.1997. "On the relation between NDVI, fractional vegetation cover, and leaf area index". *Remote Sensing Environment* 62: 241-252.

Caturegli, L, M Corniglia, M, Gaetani, N Grossi, S Magni, M Migliazzi, L Angelini, M Mazzoncini, N Silvestri, M Fontanelli, M Raffaelli, A Peruzzi and M Volterrani. 2016. "Unmanned Aerial Vehicle to Estimate Nitrogen Status of Turf grasses". *PLoS ONE* 11: 1-13.

Ceballos, G, PR Ehrlich, AD Barnosky, A Garcia, RM Pringle and TM Palmer. 2015. "Accelerated modern human-induced species losses: Entering the sixth mass extinction". *Science Advances* 1: e1400253. https://doi.org/10.1126/sciadv.1400253

Ceccato, P, S Flasse, S Tarantola, S Jacquemoud and JM Grégoire.2001. "Detecting vegetation leaf water content using reflectance in the optical domain". *Remote Sensing Environment* 77: 22-33.

Chang, A, Jinha Jung, Maeda M Murilo and Juan Landivar.2017. "Crop height monitoring with digital imagery from Unmanned Aerial System (UAS)". *Computers and Electronics in Agriculture*141: 232-237.

Chapman, SC, T Merz, A Chan. 2014. "Pheno-copter: a low-altitude, autonomous remote-sensing robotic helicopter for high-throughput field-based phenotyping". *Agronomy* 4(2): 279-301.

Chen Z, Ren J, Tang H. 2016. "Progress and perspectives on agricultural remote sensing research and applications in China". *Journal of remote sensing* 20(5): 748-67.

Chen, J, L Tang and XJ Liu. 2011. "Modeling plant nitrogen uptake and grain protein accumulation in rice". *SciAgricSin.,*44(10): 1997-2004.

Chen, LP. 2003. "Theoretical and experimental studies on variable-rate fertilization in precision farming". Beijing: China Agricultural University.

Chen, P, D Haboudane, N Tremblay. 2010. "New spectral indicator assessing the efficiency of crop nitrogen treatment in corn and wheat". *Remote Sensing of Environment* 114(9): 1987-97.

Chen, P, S Jiulin, W Jihua. 2010. "Using remote sensing technology for crop nitrogen diagnosis: status and trends". Sci Sin Inform. 40(Supp.): 21-37.

Chen, S, Yubin Lan, Zhiyan Zhou, Fan Ouyang, Guobin Wang, Xiaoyu Huang, Xiaoling Deng and Shengnan Cheng. 2020. "Effect of Droplet Size Parameters on Droplet Deposition and Drift of Aerial Spraying by Using Plant Protection UAV". *Agronomy* 10: 195. doi:10.3390/agronomy10020195

Chen, SD, YB Lan and JY Li. 2017. "Effect of wind field below rotor on distribution of aerial spraying droplet deposition by using multi-rotor UAV". *Transactions of the Chinese Society for Agricultural Machinery* 48: 105-113.

Chen, TH and SH Lu. 2012. "Autonomous navigation control system of agricultural mini-unmanned aerial vehicles based on DSP". *Transactions of the Chinese Society for Agricultural Machinery* 28: 164-169.

Cho, MA and AK Skidmore. 2006. "A new technique for extracting the red edge position from hyperspectral data: the linear extrapolation method". *Remote Sensing of Environment* 101(2): 181-93.

Chunhua, X, L Shaokun, W Keru. 2008. "Response of canopy direction reflectance spectrum for the wheat vertical leaf distribution". SciAgric Sin. 41(8): 2271-8.

Clarke, R. 2014. "Understanding the drone epidemic". 2014. *Comput Law Secur Rev* 30: 230-246.

Corti, M, D Cavalli, G Cabassi, A Vigoni, L Degano and P Marino Gallina. 2018. "Application of A Low-cost Camera on A UAV to Estimate Maize Nitrogen-related Variables". *Precision Agriculture.*

Cui, D and D Curry. 2005. "Prediction in marketing using the support vector machine". *Mark. Sci* 24: 595-615.

Cunha, M, C Carvalho and ARS Marcal. 2012. "Assessing the ability of image processing software to analyse spray quality on water-sensitive papers used as artificial targets". *Biosystems Engineering* 111(1): 11-23.

Dash, JP, S Watt Michael, DPearseGrant, Marie Heaphy andS Dungey Heidi. 2017. "Assessing very high resolution UAV imagery for monitoring forest health during a simulated disease outbreak". *ISPRS Journal of Photogrammetry and Remote Sensing* 131: 1-14.

De Castro, A, F Jiménez-Brenes, J Torres-Sánchez, J Peña, I Borra-Serrano and F López-Granados. 2018. "3-D characterization of vineyards using a novel UAV imagery-based OBIA procedure for precision viticulture applications". *Remote Sensing* 10; 584.

De Castro, AI, J Torres-Sánchez, JM Peña, FM Jiménez-Brenes, O Csillik and F López-Granados. 2018. "An Automatic Random Forest-OBIA Algorithm for Early Weed Mapping between and within Crop Rows Using UAV Imagery". *Remote Sensing* 10: 285.

De Souza, MF, Lucas Rios do Amaral, Stanley Robson de Medeiros Oliveira, Marcos Antonio Neris Coutinho and Camila Ferreira Netto. 2020. "Spectral differentiation of sugarcane from weeds". *Biosystems Engineering* 190: 41-46.

Deery, DM, GJ Rebetzke, JA Jimenez-Berni. 2016. "Methodology for high through put Field phenotyping of canopy temperature using airborne thermography". *Frontiers in Plant Science* 7(12): 1808-21.

Delalieux, S, PJ Zarco-Tejada, L Tits, JA, M Bello, DS Intrigliolo and B Somers. 2014. "Unmixing-Based Fusion of Hyperspatial and Hyperspectral Airborne Imagery for Early Detection of Vegetation Stress". *IEEE Journal of Selected Topics in Applied Earth Observations and Remote Sensing* 7: 2571-2582.

Deng, L, Z Mao, X Li, Z Hu, F Duan and Y Yan. 2018. "UAV-based multispectral remote sensing for precision agriculture: A comparison between different cameras". *ISPRS J. Photogramm. Remote Sensing* 146: 124-136.

Derksen, RC, H Zhu and HE Ozkan. 2008. "Determining the influence of spray quality, nozzle type, spray volume, and air-assisted application strategies on deposition of pesticides in soybean canopy". *American Society of Agricultural Engineers* 51: 1529-1537.

Diwate, SK, VN Nitnaware and K Argulwar. 2018. "Design and development of application specific drone machine for seed sowing". *International Research Journal of Engineering and Technology* 5 (5): 4003-4007.

Dos Santos Ferreira, A, DM Freitas, GG da Silva, H Pistori and MT Folhes. "Weed detection in soybean crops using ConvNets". *Computers and Electronics Agriculture* 143: 314-324.

Du, M and N Noguchi. 2017."Monitoring of Wheat Growth Status and Mapping of Wheat Yield's within-Field Spatial Variations Using Color Images Acquired from UAV-camera System".*Remote Sensing* 9: 289. doi:10.3390/rs9030289.

Duan, B, Shenghui Fang, Renshan Zhu, Xianting Wu, Shanqin Wang, Yan Gong and Yi Peng. 2019. "Remote Estimation of Rice Yield With Unmanned Aerial Vehicle (UAV) Data and Spectral Mixture Analysis". *Frontiers in Plant Science* 10: 204

Duan, T, SC Chapman, E Holland. 2016. "Dynamic quantification of canopy structure to characterize early plant vigour in wheat genotypes".*Journal of Experimental Botany* 67(15): 4523-34.

Dunford, R, K Michel, M Gagnage, H Piegay and ML Tremelo.2009. "Potential and constraints of unmanned aerial vehicle technology for the characterization of mediterranean riparian forest". *International Journal of Remote Sensing* 30: 4915-4935.

Dur[3]o, G, KrystynaJagie[33]o-Leñczuk, MariuszKormanek, Stanis[3]awMa[3]ek, Jacek Banach andKatarzynaDudek. 2015. "Using unmanned aerial vehicle (uav) to monitor the physiological condition of plants in a nursery". ISBN 978-80-228-2920-5

Düzgün, HS and N Demirel. 2011. "Remote Sensing of the Mine Environment; CRC Press: Boca Raton", FL, USA.

Ehlert, D, R Adamek and HJ Horn. 2009. "Laser rangefinder-based measuring of crop biomass under field conditions". *Precision Agriculture* 10(5): 395-408.

Espinoza, CZ, LR Khot, S Sankaran and PW Jacoby. 2017. "High Resolution Multispectral and Thermal Remote Sensing-Based Water Stress assessment In Subsurface Irrigated Grapevines. *Remote Sensing* 9: 961.

Esquerdo, JCDM, JZ Junior and FJG Antunes. 2011. "Use of NDVI/AVHRR time-series profiles for soybean crop monitoring in Brazil". *International Journal of Remote Sensing* 32: 3711-3727.

Evans, I, TH Jones, K Pang, MN Evans, SSaimin and B Goossens. 2015. "Use of drone technology as a tool for behavioral research: A case study of crocodilian nesting". *Herpetology Conservation and Biology*10: 90-98.

Ezat, MA, CJ Fritsch and CT Downs. 2018."Use of an unmanned aerial vehicle (drone) to survey Nile crocodile populations: A case study at Lake Nyamithi, Ndumo game reserve, South Africa". *Biological Conservation Journal* 223: 76-81.

Fan, QN. 2011. "The research on the pesticide spray system using for the mini unmanned helicopter". Nanjing Forestry University.

Fan, X, K Kawamura, TD Xuan, N Yuba, J Lim, R Yoshitoshi, TN Minh, Y Kurokawa and T Obitsu. 2018. "Low-cost visible and near-infrared camera on an unmanned aerial vehicle for assessing the herbage biomass and leaf area index in an Italian ryegrass field". *Grassland Science* 64: 145-150.

Felderhof, L and D Gillieson. 2012. "Near-infrared imagery from unmanned aerial systems and satellites can be used to specify fertilizer application rates in tree crops". *Journal of Remote Sensing* 37: 376-386.

Felderhof, L and D Gillieson. 2012. "Near-infrared imagery from unmanned aerial systems and satellites can be used to specify fertilizer application rates in tree crops". *Canadian Journal of Remote Sensing* 37(4): 376-386. DOI: 10.5589/m11-046

Feng, A, M Zhang, KA Sudduth, ED Vories and J Zhou. 2019."Cotton yield estimation from uav-based plant height". *American Society of Agricultural and Biological Engineers*62(2): 393-403. ISSN 2151-0032 https://doi.org/10.13031/trans.13067

Fernaìndez, JE, A Diaz-Espejo, JM Infante, P Duran, MJ Palomo, V Chamorro, IF Giroìn and L Villagarcia. 2006. "Water relations and gas exchange in olive trees under regulated deficit irrigation and partial root zone drying". *Plant Soil* 284(1-2): 273-291.

Francisca López-Granados, Jorge Torres-Sánchez, Ana-Isabel De Castro, Angélica Serrano-Pérez, Francisco-Javier Mesas-Carrascosa and José-Manuel Peña. 2016. "Object-based early monitoring of a grass weed in a grass crop using high resolution UAV imagery". *Agronomy for Sustainable Development* 36: 67. DOI 10.1007/s13593-016-0405-7

Franco, C, C Guada, JT Rodríguez, J Nielsen, J Rasmussen, D Gómez and J Montero. 2018. "Automatic detection of thistle-weeds in cereal crops from aerial RGB images. In International Conference on Information Processing and Management of Uncertainty in Knowledge-Based Systems". Springer: Cham, Switzerland. 441-452.

Fu, Y, G Yang, J Wang. 2014. "Winter wheat biomass estimation based on spectral indices, band depth analysis and partial least squares regression using hyperspectral measurements". *Computers and Electronics in Agriculture* 100(1): 51-59.

Fu, Z, Jie Jiang, Yang Gao, Brian Krienke, Meng Wang, KaitaiZhong,Qiang Cao, Yongchao Tian, Yan Zhu, Weixing Cao and Xiaojun Liu. 2020. "Wheat Growth Monitoring and Yield Estimation based on Multi-Rotor Unmanned Aerial Vehicle". *Remote Sensing* 12: 508. doi:10.3390/rs12030508

Fuentes, S, R De Bei, J Pech and S Tyerman. 2012. "Computational water stress indices obtained from thermal image analysis of grapevine canopies". *Irrigation Science* 30(6): 523-536.

Gabriel, JL, PJ Zarco-Tejada, PJ López-Herrera, E Pérez-Martín, M Alonso-Ayuso and M Quemada. 2017. "Airborne and Ground Level Sensors for Monitoring Nitrogen Status in A Maize Crop". *Biosystems Engineering* 160: 124-133.

Gai J, Lie Tang, Brian and L. Steward. 2019. "Automated crop plant detection based on the fusion of color and depth images for robotic weed control". *Journal of Field Robotics* 1-18.

Gamon, JA, CB Field, ML Goulden, KL Griffin, AE Hartley, G Joel, J Penuelas and R Valentini. 1995. "Relationships between NDVI, canopy structure, and photosynthesis in three Californian vegetation types". *Ecological Applications* 5: 28-41. doi:10.2307/1942049.

Gao, J, W Liao, D Nuyttens, P Lootens, J Vangeyte, A Pižurica, Y He and JG Pieters. 2018. "Fusion of pixel and object-based features for weed mapping using unmanned aerial vehicle imagery". *International Journal of Applied Earth Observation and Geoinformation* 67: 43-53.

García-Martínez, H, Héctor Flores-Magdaleno, Abdul Khalil-Gardezi, Roberto Ascencio-Hernández, Leonardo Tijerina-Chávez, Mario A Vázquez-Peña and R Oscar. 2020."Mancilla-Villa Digital Count of Corn Plants Using Images Taken by Unmanned Aerial Vehicles and Cross Correlation of Templates".*Agronomy*10: 469.

Gausman, HW, WA Allen and R Cardenas. 1969. "Reflectance of cotton leaves and their structure". *Remote Sensing Environment* 1: 19-22. doi: 10.1016/S0034-4257(69)90055-8

Geipel, J, J Link, JA Wirwahn and W Claupein. 2016. "A Programmable Aerial Multispectral Camera System for In-Season Crop Biomass and Nitrogen Content Estimation". *Agriculture* 6: 4. Doi:10.3390/agriculture6010004.

Geipel, J, Johanna Link andWilhelm Claupein. 2014."Combined Spectral and Spatial Modeling of Corn Yield Based on Aerial Images and Crop Surface Models Acquired with an Unmanned Aircraft System".*Remote Sensing* 11: 10335-10355. doi:10.3390/rs61110335

Gennaro, SFD, A Matese, B Gioli, P Toscano, A Zaldei, A Palliotti and L Genesio. 2017. "Multisensor Approach to assess Vineyard Thermal Dynamics Combining High-resolution Unmanned Aerial Vehicle (UAV) Remote Sensing and Wireless Sensor Network (WSN) Proximal Sensing". *Scientia Horticulturae* 211: 83-87.

Gholizadeh, H, AF Rahman and AF Rahman. 2015. "Comparing the Performance of Multispectral Vegetation Indices and Machine-Learning Algorithms for Remote Estimation ofChlorophyll Content: A Case Study in the Sundarbans Mangrove Forest". Abingdon: Taylor & Francis, Inc.

Girona, J, M Mata, J del Campo, A Arboneìs, E Bartra and J Marsal. 2006. "The use of midday leaf water potential for scheduling deficit irrigation in vineyards". *Irrigation Science* 24(2): 115-127. http://dx.doi.org/10.1007/s00271-005-0015-7.

Gnädinger, F and U Schmidhalter. 2017. "Digital counts of maize plants by unmanned aerial vehicles (UAVs)". *Remote Sensing* 9: 544.

Gómez-Candón, D, J Torres-Sanchez, S Labbé, A Jolivot, S Martinez and J Regnard. 2017. "Water Stress assessment at Tree Scale: High-resolution Thermal UAV Imagery Acquisition and Processing". *Acta Horticulturae* 1150: 159-166.

Gong JQ, WQ Fan and J Peng. 2019. "Application analysis of hydraulic nozzle and rotary atomization sprayer on plant protection UAV". Int J Precis Agric Aviat, 2(1): 25-29. DOI: 10.33440/j.ijpaa.20190201.0021

Gong, Y, Bo Duan, Shenghui Fang, Renshan Zhu, Xianting Wu, Yi Ma and Yi Peng. 2018."Remote estimation of rapeseed yield with unmanned aerial vehicle (UAV) imaging and spectral mixture analysis".*Plant Methods*14:70.https://doi.org/10.1186/s13007-018-0338-z

Gonzalez-Dugo, V, P Zarco-Tejada, E Nicolas, PA Nortes, JJ Alarcon, DS Intrigliolo. 2013. "Using high resolution UAV thermal imagery to assess the variability in the water status of five fruit tree species within a commercial orchard". *Precision Agriculture* 14: 660-678. doi: 10.1007/s11119-013-9322-9

Goodyer, J. 2013. "Drone rangers [africa special sustainability]". *Engineering and Technology Journal* 8: 60-61.

Gossen, BD, GaryPeng, TM Wolfand MR McDonald. 2008. "Improving spray retention to enhance the efficacy of foliar-applied disease- and pest-management products in field and row crops". *Canadian Journal of Plant Pathology* 30(4): 505-516.

Grenzdörffer, GJ, A Engelb and B Teichertc. 2008. "The photogrammetric potential of low-cost UAVs in forestry and agriculture".The International*Archives of the Photogrammetry, Remote Sensing and Spatial Information Sciences* 37(1): 1207-1214.

Gu, Y, J Yuan and C Liu. 2011. "FIS-based method to generate bivariate control parameters regulation sequence for fertilization". *Transactions of the Chinese Society Agricultural Engineering* 27(11): 134-139.

Guo, JX, K Kong and KL Xie. 2016. "Effects of nutrient management on yield and nitrogen use efficiency of direct seeding rice". *Acta Agronomica Sinica* 42(7): 1016-1025.

Guo, Q, W Yang, F Wu. 2018. "High-throughput crop phenotyping: accelerators for development of breeding and precision agriculture". *Chinese Academy of Science Bulletin* 33(9): 940-946.

Guo, S, J Li, W Yao, Y Zhan, Y Li and Y Shi. 2019. "Distribution characteristics on droplet deposition of wind field vortex formed by multirotor UAV". *PLoS ONE* 14(7): e0220024. https://doi.org/10.1371/journal.pone.0220024

GuobinWang, Yubin Lan, Haixia Qi, Pengchao Chen, Andrew Hewittd and Yuxing Hana. 2019. "Field evaluation of an unmanned aerial vehicle(UAV) sprayer: effect of spray volumeon deposition and the control of pests and disease in wheat". *Pest Management Science* 75: 1546-1555. DOI 10.1002/ps.5321

Haboudane, D, JR Miller, N Tremblay. 2002. "Integrated narrow-band vegetation indices for prediction of crop chlorophyll content for application to precision agriculture". *Remote Sensing of Environment* 81(2-3): 416-26.

Hahn, N, A Mwakatobe, J Konuche, N de Souza, J Keyyu, M Goss, A Chang'a, SPalminteri, E Dinerstein and D Olson. 2017. "Unmanned aerial vehicles mitigate human–elephant conflict on the borders of Tanzanian Parks: a case study". *Oryx* 51: 513-516.

Hahn, N, A Mwakatobe, J Konuche, N de Souza, J Keyyu, M Goss, A Chang'a, S Palminteri, E Dinerstein, and D Olson. 2017. "Unmanned aerial vehicles mitigate human–elephant conflict on the borders of Tanzanian Parks: A case study". *Oryx* 51: 513-516.

Hallett, LM, RJ Standish, J Jonson and RJ Hobbs. 2014. "Seedling emergence and summer survival after direct seeding for woodland restoration on old fields in south-western Australia". *Ecological Management Restoration* 15: 140-146.

Hambrecht, L, RP Brown, AK Piel and SA Wich. 2019. "Detecting 'poachers' with drones: Factors influencing the probability of detection with TIR and RGB imaging in miombo woodlands, Tanzania". *Biological Conservation Journal*233: 109-117.

Hamuda, E, M Glavin, and E Jones. 2016. "A survey of image processing techniques for plant extraction and segmentation in the field". *Computers and Electronics in Agriculture* 125: 184-199.

Hardwick, K, JR Healey, S Elliott and D Blakesley. 2004. "Research needs for restoring seasonal tropical forests in Thailand: Accelerated natural regeneration". *New For.*,27: 285-302.

Hassanein, M, Z Lari and N El-Sheimy. 2018. "A new vegetation segmentation approach for cropped fields based on threshold detection from hue histograms". *Sensors* 18: 1253.

Hatfield, JL, AA Gitelson, JS Schepers and CL Walthall. 2008. "Application of spectral remote sensing for agronomic decisions". *Agronomy Journal*100: 117-131.

He, ZY, Y Zhu, YD Li. 2017. "Study on estimation model for nitrogen nutrition index and yield on double cropping rice in southern China". *Journal of Nanjing Agricultural University* 40(1): 11-19.

Heidary, MA, JP Douzals, C Sinfort and A Vallet. 2014. "Influence of spray characteristics on potential spray drift of field crop sprayers: A literature review". *Crop Protection* 63(5): 120-130.

Heirloom Organics. 2018. "How to grow barley | guide to growing Barley". Retrieved August 8, from http:// www.heirloom-organics.com/guide/va/guidetogrowing barley.

Herwitz, SR, LF Johnson, RG Higgins, DV Sullivan, J Zheng and BM Lobitz. 2004. "Imaging from an unmanned aerial vehicle: agricultural surveillance and decision support". *Computers and Electronics in Agriculture* 44 (1): 49-61.

Hoffmann, H, R Jensen, A Thomsen, H Nieto, J Rasmussen and T Friborg. 2016. "Crop Water Stress Maps for an Entire Growing Season from Visible and Thermal UAV Imagery." *Biogeosciences* 13 (24): 6545-6563. doi:10.5194/bg-13-6545-2016

Holzman, ME, R Rivas and MC Piccolo. 2014. "Estimating soil moisture and the relationship with crop yield using surface temperature and vegetation index". *International Journal of Applied Earth Observation and Geoinformation* 28: 181-192. doi: 10.1016/j.jag.2013. 12.006

https://krishijagran.com/farm-mechanization/drone-that-can-plant-10-000-trees-a-day/

https://www.dronethusiast.com/history-of-drones/

https://www.google.com/amp/s/m.timesofindia.com/city/coimbatore/reaping-rich-dividends-with-drones/amp_articleshow/71382745.cms

https://www.newindianexpress.com/cities/hyderabad/2020/jan/19/spraying-pesticides-using-drones-illegal-agriculture-ministry-to-hyderabad-activist-2091512.html

https://www.thehindu.com/news/national/modi-calls-for-evergreen revolution/ article18510225.ece

Hu, P, SC Chapman, X Wang. 2018. "Estimation of plant height using a high through put phenotyping platform based on unmanned aerial vehicle and self-calibration: Example for sorghum breeding". *European Journal of Agronomy* 95(4): 24-32.

Hu, P. 2018. "High-throughput field morphological phenotyping using UAV based proximal sensing". Beijing: China Agricultural University.

Huang, Y, KN Reddy, RS Fletcher and D Pennington. 2018. "UAV low-altitude remote sensing for precision weed management". *Weed Technology* 32: 2-6.

Huang, Y, WC Hoffmann, Y Lan, W Wu and BK Fritz. 2009. "Development of a spray system for an unmanned aerial vehicle platform". *Applied Engineering in Agriculture* 25 (6): 803-809.

Huang, Yanbo, Wesley C Hoffmann, Yubin Lan, Wenfu Wu and Bradley K Fritz. 2009. "Development of a spray system for an unmanned aerial vehicle platform" (2009). Publications from USDA-ARS / UNL Faculty. 2089. https://digitalcommons.unl.edu/usdaarsfacpub/2089

Huang, YB. 2010. "Airborne remote sensing for precision aerial application of crop production and protection materials". Proceedings of 2010 Sino-US Workshop on Intelligent Equipment for Precision Agriculture and Remote Sensing in Agriculture. College Station, Texas, USA, 22 June 2010; 109-111.

Huilin Tao, Haikuan Feng, Liangji Xu, Mengke Miao, Guijun Yang, Xiaodong Yang and Lingling Fan. 2020."Estimation of the Yield and Plant Height of Winter Wheat Using UAV-Based Hyperspectral Images". *Sensors*20: 1231. doi:10.3390/s20041231

Hunt, DR. 1983. "Farm Power and Machinery Management" (8th Edition). Iowa State University Press, Ames, IA, pp. 365.

Hunt, E, Jr, D Horneck, P Hamm, D Gadler,A Bruce, R Turner, C Spinelli and J Brungardt. 2014. "Detection of nitrogen deficiency in potatoes using small unmanned aircraft systems". In Proceedings of the 12thInternational Conference on Precision Agriculture, Sacramento, CA, USA, 20-23 July 2014.

Hunt, ER, DA Horneck, CB Spinelli, RW Turner, AE Bruce, DJ Gadler, JJ Brungardt and PB Hamm. 2018. "Monitoring Nitrogen Status of Potatoes Using Small Unmanned Aerial Vehicles". *Precision Agriculture* 19: 314-333.

Hunt, ER, Jr, M Cavigelli, CST Daughtry, JE Mcmurtrey and CL Walthall. 2015. "Evaluation of Digital Photography From Model Aircraft for Remote Sensing of Crop Biomass and Nitrogen Status". *Precision Agriculture* 6: 359-378.

Hunt, ER, Jr, SI Rondon, PB Hamm, RW Turner, AE Bruce and JJ Brungardt. 2016. "Insect Detection and Nitrogen Management for Irrigated Potatoes Using Remote Sensing From Small Unmanned Aircraft Systems". In Autonomous Air and Ground Sensing Systems for Agricultural Optimization and Phenotyping. *International Society for Optics and Photonics*: Washington DC, USA, Volume 9866.

Hunt, JER, WD Hively, SJ Fujikawa, DS Linden, CST Daughtry and GW McCarty. 2010. "Acquisition of NIR-Green-Blue Digital Photographs from Unmanned Aircraft for Crop Monitoring". *Remote Sensing* 2: 290-305.

Hunter, JE, TW Gannon, RJ Richardson, FH Yelverton and RG Leon. 2019. "Coverage and drift potential associated with nozzle and speed selection for herbicide applications using an unmanned aerial sprayer". *Weed Technology*. doi: 10.1017/wet.2019.101

Hussey, NE, ST Kessel, KAarestrup, SJ Cooke, PD Cowley, AT Fisk and FG Whoriskey. 2015. "Aquatic animal telemetry: A panoramic window into the underwater world". *Science* 348: 1255642. https://doi. org/10.1126/science.1255642

Jackson, RD and AR Huete. 1991. "Interpreting vegetation indices". *Preventive Veterinary Medicine*11: 185-200.

James, JJ, TJ Svejcar and MJ Rinella. 2011. "Demographic processes limiting seedling recruitment in arid grassland restoration". *Journal of Applied Ecology* 48: 961-969.

Jay, S, F Baret, D Dutartre, G Malatesta, S Héno, A Comar, M Weiss and F Maupas. 2018. "Exploiting The Centimeter Resolution of UAV Multispectral Imagery to Improve Remote-sensing Estimates of Canopy Structure and Biochemistry In Sugar Beet Crops". *Remote Sensing of Environ*ment.

Jian Zhang, Chufeng Wang, Chenghai Yang, Tianjin Xie, Zhao Jiang, Tao Hu, Zhibang Luo, Guangsheng Zhou and Jing Xie. 2020."Assessing the Effect of Real Spatial Resolution of In-Situ UAV Multispectral Images on Seedling Rapeseed Growth Monitoring". Remote Sensing 12: 1207. doi:10.3390/rs12071207

Jiang, J, Z Zhang, Q Cao. 2019. "Use of a digital camera mounted on a consumer-grade unmanned aerial vehicle to monitor the growth status of wheat". *Journal of Nanjing Agricultural University* 42(4): 622-31.

Jingyao Gai, Lie Tang, Brian L Steward. 2019. "Automated crop plant detection based on the fusion of color and depth images for robotic weed control". *Journal of Field Robotics* 1-8. DOI: 10.1002/rob.21897

Johansen, K, MJL Morton, Y Malbeteau, B Aragon, S Al-Mashharawi, M Ziliani, Y Angel, G Fiene, S Negrao, MAA Mousa, MA Tester and MFMccabe.2019. "Predicting biomass and yield at harvest of salt-stressed tomato plants using uav imagery".*The International Archives of the Photogrammetry, Remote Sensing and Spatial Information Sciences* Volume XLII-2/W13, ISPRS Geospatial Week 2019, 10-14 June 2019, Enschede, The Netherlands.

Jonckheere, I, S Fleck, K Nackaerts. 2004. "Review of methods for in situ leaf area index determination: Part I. Theories, sensors and hemispherical photography". *Agriculture and Forest Meteorology* 121(1-2): 19-35.

Jones, HG, M Stoll, T Santos, C de Sousa, MM Chaves and OM Grant. 2002. "Use of infrared thermography for monitoring stomatal closure in the field: application to grapevine". *Journal of Experimental Botany* 53(378): 2249–2260. http://dx.doi.org/10.1093/jxb/erf083. PubMed

Jorge, J, M Vallbé and JA Soler. 2019. "Detection of irrigation inhomogeneities in an olive grove using the NDRE vegetation index obtained from UAV images". *European Journal of Remote Sensing* 52: 169-177.

Jureèka, F, Vojtìch Lukas, Petr Hlavinka, Daniela Semerádová, ZdenìkŽalud and Miroslav Trnka. 2018. "Estimating crop yields at the field level usingLandsat and modis products". Actauniversitatisagriculturae et silviculturaemendelianaebrunensis66: 1141-1150.

K¡rížová, K, M Kroulík, J Haberle, J Lukáš and J Kumhálová. 2018. "Assessment of soil electrical conductivity using remotely sensed thermal data". *Agronomy Research* 16: 784-793.

Kamilaris, A and FX Prenafeta-Boldu. 2018. "Deep learning in agriculture: A survey". *Computers and Electronics in Agriculture* 147: 70-90.

Karanth, KK, AM Gopalaswamy, PK Prasad and S Dasgupta. 2013. "Patterns of human-wildlife conflicts and compensation: Insights fromWestern Ghats protected areas". *Biological Conservation Journal*166: 175-185.

Karcher, DE and MD Richardson. 2003. "Quantifying turfgrass color using digital image Analysis". *Crop Sciences* 43: 943-951.

Katsigiannis, P, L Misopolinos, V Liakopoulos, TK Alexandridis and G Zalidis. 2016. "An Autonomous Multi-sensor UAVSystem for Reduced-input Precision Agriculture Applications". In Proceedings of the 2016 24th Mediterranean Conference on Control and Automation (MED), Athens, Greece, 21-24 June 2016; pp. 60-64.

Kays, R, MC Crofoot, W Jetz and MWikelski. 2015. "Terrestrial animal tracking as an eye on life and planet". *Science* 348:aaa2478. https://doi. org/10.1126/science.aaa2478

Kerkech, M, A Hafiane and R Canals. 2018. "Deep leaning approach with colorimetric spaces and vegetation indices for vine diseases detection in UAV images". *Computers and Electronics Agriculture* 155: 237-243.

Khan, Z, V Rahimi-Eichi, S Haefele, T Garnett and SJ Miklavcic. 2018. "Estimation of vegetation indices forhigh-throughput phenotyping of wheat using aerial imaging". *Plant Methods* 14: 20.

Kim, DM, H Zhang, H Zhou, Dut, Q Wu, TCMockler and MY Berezin.2015. "Highly sensitive image-derived indices of water-stressed plants using hyperspectral imaging in SWIR and histogram analysis". *Scientific Rep*orts 515919: 1-11.doi: 10.1038/srep15919.

Kirk, IW. 2004. "Aerial Applicators Spray Nozzle Handbook". AH-726. Washington, D.C., USDA Agricultural Research Service.

Kitchen, NR, KA Sudduth, ST Drummond. 2010. "Ground-based canopy reflectance sensing for variable-rate nitrogen corn fertilization". *Agronomy Journal* 102(1): 71-84.

Knoth, C, B Klein, T Prinz and T Kleinebecker. 2013. "Unmanned aerial vehicles as innovative remote sensing platforms for high-resolution infrared imagery to support restoration monitoring in cut-over bogs". *Applied Vegetation Science* 16: 509-517.

Koh, LP and SA Wich. 2012."Dawn of drone ecology: Low-cost autonomous aerial vehicles for conservation". *Tropical Conservation Science*5: 121-132.

Konishi, N, H Shiga, D Asaka. 2000. "The application of remote sensing technology in high quality rice production". Measurement. 10(7): 14-20.

Krienke, B, RB Ferguson, M Schlemmer, K Holland. 2017. "Using an unmanned aerial vehicle to evaluate nitrogen variability and height effect with an active crop canopy sensor". *Precision Agriculture* 18(6): 900-15.

Lan, Y, X Deng and G Zeng. 2019. "Advances in diagnosis of crop diseases, pests and weeds by UAV remote sensing". *Smart Agriculture* 1(2): 1-19.

Lan, YB, SJ Thomson, YB Huang, WC Hoffmann and HH Zhang. 2010. "Current status and future directions of precision aerial application for site-specific crop management in the USA". *Computers and Electronics in Agriculture* 74: 34-38.

Lan, YB, WC Hoffmann, BK Fritz, DE Martin, JD Lopez and Jr. 2008. "Spray drift mitigation with spray mix adjuvants". *Applied Engineering in Agriculture* 24: 5-10.

Landis, JR and GG Koch. 1977. "The measurement of observer agreement for categorical data". *Biometrics* 33(1): 159-174.

Laursen, MS, RN Jørgensen, HS Midtiby, K Jensen, MP Christiansen, TM Giselsson, AK Mortensen and PK Jensen. 2016. "Dicotyledon Weed Quantification Algorithm for Selective Herbicide Application in Maize Crops". *Sensors* 16: 1848.

Lee, W and SW Searcy. 2000. "Multispectral sensor for detecting nitrogen in corn Plants". In: 2000 ASAE annual international meeting sponsored by ASAE, July 9-12.

Lehmann, JRK, T Prinz, SR Ziller, J Thiele, G Heringer, JAA Meira-Neto and TK Buttschardt. 2017. "Open-Source Processing and Analysis of Aerial Imagery Acquired with a Low-Cost Unmanned Aerial System to Support Invasive Plant Management". *Frontiers in Environmental Science* 5.

Leica Geosystems. ERDAS Imagine; Leica Geosystems: Atalanta, GA, USA, 2004.

Lelong, CCD, P Burger, G Jubelin, B Roux and SLF Baret. 2008. "Assessment of Unmanned Aerial Vehicles Imagery for Quantitative Monitoring of Wheat Crop In Small Plots". *Sensors* 8: 3557-3585.

Li, F, Y Miao, G Feng, F Yuan, S Yue, X Gao. 2014. "Improving estimation of summer maize nitrogen status with red edge-based spectral vegetation indices". *Field Crops Research* 157: 111-123.

Li, H, C Zhao, W Huang. 2013. "Non-uniform vertical nitrogen distribution within plant canopy and its estimation by remote sensing: a review". *Field Crops Research* 142(2): 75-84.

Li, J, F Zhang, X Qian, Y Zhu and G Shen. 2015. "Quantification of Rice Canopy Nitrogen Balance Index with Digital Imagery From Unmanned Aerial Vehicle". *Remote Sensing Letters* 6: 183-189.

Li, L, Y Fan, X Huang and L Tian. 2016. "Real-time UAV Weed Scout for Selective Weed Control By Adaptive Robust Control and Machine Learning Algorithm". In Proceedings of the 2016 American Society of Agricultural and Biological Engineers Annual International Meeting, as ABE 2016, Orlando, FL, USA, 17–20 July 2016; American Society of Agricultural and Biological Engineers: St. Joseph, MO, USA.

Li, P and Q Wang. 2011. "Retrieval of leaf biochemical parameters using PROSPECT Inversion: a new approach for alleviating Ill-posed problems". *IEEE Transaction on Geoscience and Remote Sensing* 49(7): 2499-506.

Li, R, Y Ding and H Yu. 2017. "Construction and application of the electronic prescription map system of variable rate fertilization in Southern Jiangsu". *Journal of China Agricultural University* 22(8): 124-32.

Li, W, H Li, J Wang. 2010. "A study on classification and monitoring of winter wheat growth status by Landsat/TM image". *Journal of Triticeae Crops.* 30(1): 92-95.

Li, W, J Wang, C Zhao. 2008. "A model for predicting protein content in winter wheat grain based on Land-Sat TM image and nitrogen accumulation". *International journal of remote sensing* 12(3): 506-14.

Li, Y, S Shu, L Chen. 2013. "Photosynthetic production simulation based on plant type in rice: a review". *Chinese Agricultural Science Bulletin* 29(33): 1-5.

Liakos, K, P Busato, D Moshou, S Pearson and D Bochtis. 2018. "Machine learning in agriculture: A review". *Sensors* 18: 2674.

Liang, T, Z Xiangcheng, C Mengying. 2012. "Relationships of rice canopy PAR interception and light use efficiency to grain yield". *Chinese Journal of Applied Ecology* 23(5): 1269-76.

Linchant, J, J Lisein, J Semeki, P Lejeune and C Vermeulen. 2015. "Are unmanned air- craft systems (UASs) the future of wildlife monitoring? A review of accomplishments and challenges". *Mammal Review* 45: 239-252.

Linchant, J, J Lisein, J Semeki, P Lejeune and CVermeulen. 2015."Are unmanned aircraft systems (UAS s) the future of wildlife monitoring? A review of accomplishments and challenges". *Mammal Review*45: 239-252.

Liu, BH, XP Chen and ZL Cui. 2015. "Research advance in yield potential and yield gap of three major cereal crops". *Chinese Journal of Eco-Agriculture* 23(5): 525-534.

Liu, C, Z Wang, Z Chen. 2018. "Nitrogen monitoring of winter wheat based on unmanned aerial vehicle remote sensing image". *Transactions Chinese Society of Agricultural Machinery* 49(6): 207-14.

Liu, H, H Zhu and P Wang. 2017. "Quantitative Modelling for Leaf Nitrogen Content of Winter Wheat Using UAV-based Hyperspectral Data". *International Journal of Remote Sensing* 38: 2117-2134.

Liu, H, L Tang, W Zhang. 2009. "Construction and implementation of model-based visual rice growth system". *Transactions Chinese Society of Agricultural Engineering* 25(9): 148-54.

Liu, L. 2002. "Hyperspectral remote sensing application in precision agriculture". Beijing: Post-doctoral Dissertation of the Chinese Academy of Sciences.

Liu, S, L Li, W Gao, Y Zhang, Y Liu, S Wang and J Lu. 2018. "Diagnosis of Nitrogen Status In Winter Oilseed Rape (*Brassic anapus* L.) Using In-situ Hyperspectral Data and Unmanned Aerial Vehicle (UAV) Multispectral Images". *Computers and Electronic Agriculture* 151: 185-195.

Liu, XJ, YC Tian, X Yao. 2012. "Monitoring leaf water content based on hyperspectra in rice". SciAgric Sin. 45(3): 435-42.

López-Bao, JV, J Frank, L Svensson, M Akesson and ALangefors. 2017. "Building public trust in compensation programs through accuracy assessments of damage verification protocols". *Biological Conservation Journal* 213: 36-41.

López-Granados, F, J Torres-Sánchez, A Serrano-Pérez, AI de Castro, FJ Mesas-Carrascosa, and JM Peña. 2015. "Early season weed mapping in sunflower using UAV technology: Variability of herbicide treatment maps against weed thresholds". *Precision Agriculture* 17: 183-199.

Lu, N, W Wang, Q Zhang, D Li, X Yao, Y Tian, Y Zhu, W Cao, F Baret, S Liu and T Cheng. 2019. "Estimation of Nitrogen Nutrition Status in Winter Wheat From Unmanned Aerial Vehicle Based Multi-Angular Multispectral Imagery". *Frontiers Plant Science* 10: 1601. Doi: 10.3389/fpls.2019.01601

Lukina, EV, KW Freeman, KJ Wynn. 2001. "Nitrogen fertilization optimization algorithm based on in-season estimates of yield and plant nitrogen uptake". *Journal of Plant Nutrition* 24(6): 885-98.

Lv, MQ, SP Xiao, Y Tang and Y He. 2019. "Influence of UAV flight speed on droplet deposition characteristics with the application of infrared thermal imaging". *International Journal of Agriculture and Biological Engineering* 12(3): 10-17.

Ma, J, Y Zhu Y, X Yao. 2007. "Relationship between leaf nitrogen content and fluorescence parameters in rice". *Chinese Journal of Rice Science* 21(1): 65-70.

Ma, Y, X Song, B Quan. 2015. "Temporal and spatial variation of plant nitrogen and biomass of winter wheat under variable rate fertilization conditions". *Journal of Triticeae Crops* 35(11): 1576-87.

Maes, WH and K Steppe. 2019. "Perspectives for remote sensing with unmanned aerial vehicles in precision agriculture". *Trends in Plant Science* 24(2): 152-64.

Maimaitijiang, M, A Ghulam, P Sidike, S Hartling, M Maimaitiyiming, K Peterson, E Shavers, J Fishman, J Peterson, S Kadam. 2017. "Unmanned Aerial System(UAS) based Phenotyping of Soybean Using Multi-sensor Data Fusion and Extreme Learning Machine". *ISPRS Journal of Photogrammetry and Remote Sensing* 134: 43-58.

Mäkynen, J, H Saari, C Holmlund, R Mannila and T Antila.2012. "Multi- and hyperspectral UAV imaging system for forest and agriculture applications". Next-generation spectroscopic technologies. Druy M.A., Crocombe R.A., (Eds.) Proc. SPIE: Baltimore, Maryland, USA 8374: 1-9.

Maleki, MR, AM Mouazen, H Ramon. 2007. "Multiplicative scatter correction during on-line measurement with near infrared spectroscopy." *Biosystems Engineerig* 96(3): 427-433.

Malenovsky, Z, A Lucieer, DH King, JD Turnbull and SA Robinson. 2017. "Unmanned Aircraft System Advances Health Mapping of Fragile Polar Vegetation". *Methods in Ecology and Evolution* 8: 1842-1857.

Marcial-Pablo, MdJ, A Gonzalez-Sanchez, SI Jimenez-Jimenez, RE Ontiveros-Capurata and W Ojeda-Bustamante. 2019. "Estimation of vegetation fraction using RGB and multispectral images from UAV". *International journal of remote sensing* 40: 420-438.

Marino, S and A Alvino. 2018. "Detection of homogeneous wheat areas using multi-temporal UAS images and ground truth data analyzed by cluster analysis". *European Journal of Remote Sensing* 51: 266-275.

Markle, JC, FJ Niederholzer and FG Zalom. 2016. "Evaluation of spray application methods for navel orange worm control in almonds". *Pest Management Science* 72(12): 2339-2346.

Martýìn-Vertedor, AI, JM PeìrezRodrýìguez, HP Losada and EF Castiel. 2011. "Interactive responses to water deficits and crop load in olive (*Olea europaea* L., cv. Morisca) Growth and water relations". *Agricultural Water Management* 98 (6): 941-949.

McConnell, T, E Johnson and B Burns. 2000. "A Guide to Conducting Aerial Sketchmap Surveys". US Forest Service, fhtet 00-01. Forest Health Technology Enterprise Team, Fort Collins, CO. 88.

McDonald, AW, JP Bakker and K Vegelin. 2018. "Seed bank classification and its importance for the restoration of species-rich flood-meadows". *Journal of Vegetation Science* 7: 157-164.

Medrano, H, JM Escalona, J Bota, J Gulýìas and J Flexas. 2002. "Regulation of photosynthesis of C3 plants in response to progressive drought: stomatal conductance as a reference parameter". *Annals of Botany* 89 (7): 895-905. http://dx.doi.org/10.1093/aob/mcf079. PubMed

Meiqiao, QLv, SP Xiao, Y Tang and Y He. 2019. "Influence of UAV flight speed on droplet deposition characteristics with the application of infrared thermal imaging". *International Journal of Agriculture and Biological Engineering* 12(3): 10-17. DOI: 10.25165/j.ijabe.20191203.4868

Meng, YH, YB Lan, GY Mei, YW Guo, JL Song and ZG Wang. 2018. "Effect of aerial spray adjuvant applying on the efficiency of small unmanned aerial vehicle on wheat aphids". *International Journal of Agriculture and Biological Engineering* 11(5): 46-53.

Migdall, S, H Bach, J Bobert, M Wehrhan and W Mauser. 2009. "Inversion of a canopy reflectance model using hyperspectral imagery for monitoring wheat growth and estimating yield". *Precision Agriculture* 10: 508-524. http://dx.doi.org/10.1007/ s11119-009-9104-6.

Montero, D and C Rueda. 2018. "Detection of palm oil bud rot employing artificial vision. In IOP Conference Series: Materials Science and Engineering". IOP Publishing: Bristol, UK, 437: 012004.

Moran, MS, TR Clarke, Y Inoue and A Vidal. 1994. "Estimating crop water deficit using the relation between surface-air temperature and spectral vegetation index". *Remote Sensing Environment* 49(3): 246-263. doi:10.1016/0034-4257(94)90020-5.

Moreno-Opo, R and J Pique. 2018."Reconciling the conservation of the purple swamphen (*Porphyrioporphyrio*)and its damage in Mediterranean rice fields through sustainable non-lethal techniques". *Peer reviewed Journal* 1-19.

Mozgeris, G, D Jonikavi¡cius, D Jovarauskas, R Zinkevi¡cius, S Petkevi¡cius and D Steponavi¡cius. 2018. "Imaging from manned ultra-light and unmanned aerial vehicles for estimating properties of spring wheat". *Precision Agriculture* 19: 876-894.

Mulero-Pázmány, M, R Stolper, LD Van Essen, JJ Negro and T Sassen. 2014. "Remotely piloted aircraft systems as a rhinoceros anti-poaching tool in Africa". *PLoS ONE* 9: e83873.

MuleroPázmány, M. 2015."Unmanned Aerial Systems in Conservation Biology". Ph.D. Thesis, Universidad de Sevilla, Sevilla, Spain.

Mulla, DJ. 2013. "Twentyfive years of remote sensing in precision agriculture: key advances and remaining knowledge gaps". *Biosystems Engineering* 114: 358-371.

Näsi, R, N Viljanen, J Kaivosoja, K Alhonoja, T Hakala, L Markelin and E Honkavaara. 2018. "Estimating Biomass and Nitrogen Amount of Barley and Grass Using

UAV and Aircraft Based Spectral and Photogrammetric 3D Features". *Remote Sensing* 10: 1082.

Nebiker, S, N Lack, M Abächerli and S Läderach. 2016."Light-weight multispectral uav sensors and their capabilities for predicting grain yield and detecting plant diseases".The International Archives of the Photogrammetry, Remote Sensing and Spatial Information Sciences, Volume XLI-B1, XXIII ISPRS Congress, 12-19 July 2016, Prague, Czech Republic.

Nevill, P, AT Cross and KW Dixon. 2018. "Ethical sourcing of wild seeds - a key issue in meeting global restoration targets". *Current Biology* 28: R1378-R1379.

Nguy-Robertson, AL, Y Peng, AA Gitelson. 2014. "Estimating green LAI in four crops: potential of determining optimal spectral bands for a universal algorithm". *Agriculture and Forest Meteorology* 192-193(7): 140-148.

Ni, J, TT Wang, X Yao. 2013. "Design and experiments of multi-spectral sensor for rice and wheat growth information". *Transactions Chinese Society of Agricultural Machinery* 44(5): 207-212.

Ni, J, X Yao, Y Tian. 2013. "Design and experiments of portable apparatus for plant growth monitoring and diagnosis". *Transactions Chinese Society of Agricultural Engineering*

Nishar, A, S Richards, D Breen, J Robertson and B Breen. 2016. "Thermal infrared imaging of geothermal environments and by an unmanned aerial vehicle (UAV): A case study of the Wairakei – Tauhara geothermal field, Taupo, New Zealand". *Renewable Energy* 86: 1256-1264.

Noland, RL, M Scott Wellsa, ACoulteraJeffrey, Tyler Tiedea, MBakera John,LMartinsonaKrishona, RL Craig C. Sheaffe. 2018."Estimating alfalfa yield and nutritive value using remote sensing and air temperature".*Field Crops Research* 222: 189-196

Norasma, CYN, MY Abu Sari, MA Fadzilah, MR Ismail, MH Omar, B Zulkarami, YMM Hassim and Z Tarmidi. 2018. "Rice crop monitoring using multirotor UAV and RGB digital camera at early stage of growth". IOP Conference Series: Earth and Environmental Science 169: 012095.

Ogden, LE. 2013.Drone ecology. BioScience.

Palossi, D, A Gomez, S Draskovic, K Keller, L Benini and L Thiele. 2018. "Self-sustainability in nano unmanned aerial vehicles: A blimp case study". In Proceedings of the Computing Frontiers Conference, Siena, Italy, 15–17 May 2017.

Park, S, D Ryu, S Fuentes, H Chung, E Hernández-Montes and M O'Connell. 2017. "Adaptive Estimation of Crop Water Stress in Nectarine and Peach Orchards Using High-Resolution Imagery from an Unmanned Aerial Vehicle (UAV)". *Remote Sensing* 9: 828

Pascuzzi, S, AS Anifantis, V Cimino and F Santoro. 2018. "Unmanned aerial vehicle used for remote sensing on an apulian farm in southern Italy". In Proceedings of the 17th International Scientific Conference Engineering for Rural Development, Jelgava, Latvia, 23-25 May 2018; 23-25.

Peng, S, FV Garcia, RC Laza. 996. "Increased N-use efficiency using a chlorophyll meter on high-yielding irrigated rice". *Field Crops Research* 47(23): 243-52.

Peng, Y, AA Gitelson and T Sakamoto. 2013. "Remote estimation of gross primary productivity in crops using MODIS 250m data". *Remote Sensing Environment* 128: 186-196. doi: 10.1016/j.rse.2012.10.005

Pérez-Ortiz, M, JM Peña, PA Gutiérrez, J Torres-Sánchez, C Hervás-Martínez and F López-Granados. 2016. "Selecting patterns and features for between- and within- crop-row weed mapping using UAV-imagery". *Expert Systems with Applications* 47: 85-94.

Pimm, SL, S Alibhai, R Bergl, A Dehgan, C Giri, Z Jewell, L Joppa, R Kays and S Loarie. 2015."Emerging technologies to conserve biodiversity". *Trends in Ecology and Evolution* 30: 685-696.

Pittman, JJ, DB Arnall, SM Interrante, CA Moffet and TJ Butler. 2015. "Estimation of biomass and canopy height in bermudagrass alfalfa, and wheat using ultrasonic, laser, and spectral sensors". *Sensors* 15: 2920-2943.

Poblete, T, S Ortega-Farías, M Moreno and M Bardeen. 2017. "Artificial neural network to predict vine water status spatial variability using multispectral information obtained from an unmanned aerial vehicle (UAV)". *Sensors* 17: 2488.

Poblete-Echeverrýìa, C, D, Sepulveda-Reyes, S Ortega-Farias, M ZunÞiga and S Fuentes. 2016. "Plant water stress detection based on aerial and terrestrial infrared thermography: a study case from vineyard and olive orchard". ActaHortic. 1112. XXIX IHC – Proc. Int. Symp. on Water, Eco-Efficiency and Transformation of Organic Waste in Horticultural Production pg.,l141-146

Poenaru, V, A Badea, SMCimpeanu and A Irimescu. 2015. "Multi-temporal multi-spectral and radar remote sensing for agricultural monitoring in the braila plain". *Agriculture and Agricultural Science Procedia* 6: 506-516.

Pudelko, R, T Stuczynski and M Borzecka-Walker. 2012. "The suitability of an unmanned aerial vehicle (UAV) for the evaluation of experimental fields and crops". *Agriculture* 99 (4): 431-436.

Qaio, X, X Ma, XC Zhang. 2008. "Response of coronary spectrum on chlorophyll and K information of soy". *Transactions Chinese Society of Agricultural Machinery* 39(4): 108-11.

Qiao, X. 2008. Research of soybean nutrition information diagnosis based on spectrum technology and system of variable rate fertilizer. Changchun: Jilin University.

Qin, WC, BJ Qiu and XY Xue. 2016. "Droplet deposition and control effect of insecticides sprayed with an unmanned aerial vehicle against plant hoppers". *Crop Protection* 85: 79-88.

Qin, WC, XY Xue and LX Zhou. 2014. "Effects of spraying parameters of unmanned aerial vehicle on droplets deposition distribution of maize canopies'. *Transactions of the Chinese Society for Agricultural Machinery* 30: 50-56.

Qin, WC, XY Xue, LX Zhou, SC Zhang, Z Sun, W Kong. 2014. "Effects of spraying parameters of unmanned aerial vehicle on droplets deposition distribution of maize canopies". *Transactions of the Chinese Society of Agricultural Engineering* 30(5): 50-56.

Qin, WC, XY Xue, SM Zhang, W Gu and BK Wang. 2018. "Droplet deposition and efficiency of fungicides sprayed with small UAV against wheat powdery mildew". *International Journal of Agriculture and Biological Engineering* 11(2): 27-32. DOI. 10.25165/j.ijabe.20181102.3157

Quebrajo, L, M Perez-Ruiz, L Pérez-Urrestarazu, G Martínez and G Egea.2018. "Linking thermal imaging and soil remote sensing to enhance irrigation management of sugar beet". *Biosystems Engineering* 165: 77–87.

Raeva, PL, J Šedina and A Dlesk. 2019. "Monitoring of crop fields using multispectral and thermal imagery from UAV". *European Journal of Remote Sensing* 52: 192-201.

Raeva, PL, Jaroslav Šedina and Adam Dlesk. 2019."Monitoring of crop fields using multispectral and thermal imagery from UAV".*European Journal of Remote Sensing* 52(1): 192-201. DOI: 10.1080/22797254.2018.1527661.

Rahman, A, K Khan, NY Krakauer. 2012. "Use ofremote sensing data for estimation of aman rice yield". Int. J. Agric. For. 2: 101-107. doi: 10.5923/j.ijaf. 20120201.16

Rajan, N and SJ Maas. 2009. "Mapping crop ground cover using airborne multispectral digital imagery". *Precision Agriculture* 10(4): 304-18.

Rasmussen, J, J Nielsen, F Garcia-Ruiz, S Christensen and JC Streibig. 2013. "Potential Uses of Small Unmanned Aircraft Systems (UAS) In Weed Research". *Weed Research* 53: 242-248.

Raun, WR, JB Solie and KL Martin. 2005. "Growth stage, development, and sapatial variability in corn evaluated using optical sensor readings". *Journal of Plant Nutrition* 28(1): 173-82.

Rebetzke, GJ, JA Jimenez-Berni, WD Bovill. 2016. "High-throughput phenotyping technologies allow accurate selection of stay-green". *Journal of Experimental Botany* 67(17): 4919-24.

Redpath, SM, J Young, A Evely, WM Adams, WJ Sutherland, A Whitehouse, A Amar, RA Lambert, JDC Linnell, A Watt and RJ Gutiérrez. 2013. "Understanding and managing conservation conflicts". *Trends in Ecology and Evolution*28: 100-109.

Rey, N, M Volpi, SJoost and D Tuia. 2017."Detecting animals in African Savanna with UAVs and the crowds". *Remote Sensing of Environment*200: 341-351.

Romero, M, Y Luo, B Su and S Fuentes. 2018. "Vineyard water status estimation using multispectral imagery from an UAV platform and machine learning algorithms for irrigation scheduling management". *Computers and Electronics in Agriculture* 147: 109-117.

Ronchetti, G, D Pagliari and G Sona. 2018. "DTM Generation Through UAV Survey With a FISHEYE Camera On a Vineyard". *International Archives of the Photogrammetry, Remote Sensing Spatial Information Sciences* 42: 2.

Rouse, J, R Haas, J Schell and D Deering. 1973. "Monitoring Vegetation Systems in the Great Plains with ERTS". Third ERTS Symp. NASA 309-317 SP-351.

Rowcliffe, JM and C Carbone. 2008. "Surveys using camera traps: Are we looking to a brighter future"?*Animal Conservation*11: 185-186.

Roy, M. 2009. "Agriculture in the vedic period". *Indian Journal of History of Science* 44.4: 497-520.

Sa, I, M Popovi´c, R Khanna, Z Chen, P Lottes, F Liebisch, J Nieto, C Stachniss, A Walter and R Siegwart. 2018. "Weedmap: A large-scale semantic weed mapping framework using aerial multispectral imaging and deep neural network for precision farming". *Remote Sensing* 10; 1423.

Saberioon, MM and A Gholizadeh. 2016. "Novel Approach for Estimating Nitrogen Contant In Paddy Fields Using Low Altitude Remote Sensing System". *International Archives of the Photogrammetry, Remote Sensing Spatial Information Sciences XLI-B1*, 1011–1015.

Sakamoto, T, AA Gitelson and TJ Arkebauer. 2014. "Near real-time prediction of US corn yields based on time-series MODIS data". *Remote Sensing Environment* 147: 219-231. doi: 10.1016/j.rse.2014.03.008

Sanchavat, H, H Chudhary, G Bhautik and SN Singh. 2018. "Field Evaluation of a Tractor Mounted Boom Sprayer"

Sankaran, S, LR Khot, CZ Espinoza. 2015. "Low-altitude, high-resolution aerial imaging systems for row and field crop phenotyping: a review". *European Journal of Agronomy* 70(8): 112-23.

Santesteban, LG, C Miranda and JB Royo. 2011. "Regulated deficit irrigation effects on growth, yield, grape quality and individual anthocyanin composition in *Vitis vinifera* L. cv. 'Tempranillo'". *Agricultural Water Management* 98(7): 1171–1179.

Scharf, PC and JA Lory. 2002. "Calibrating corn color from aerial photographs to Predictside dress nitrogen need". *Agronomy Journal* 94(3): 397-404.

Schirrmann, M, A Giebel, F Gleiniger, M Pflanz, J Lentschke and KH Dammer. 2016. "Monitoring agronomic parameters of winter wheat crops with low-cost UAV imagery". *Remote Sens*ing 8(9): 706.

Schneider, CA, WS Rasband and KW Eliceiri. 2012. "NIH image to Image J: 25 years of image

Shamshiri, RR, IA Hameed, SK Balasundram, D Ahmad, C Weltzien and M Yamin. 2018. "Fundamental research on unmanned aerial vehicles to support precision agriculture in oil palm plantations. *Agricultural Robots: Fundamentals and Applications* DOI:10.5772/intechopen.80936

Sharma, B and GL Ritchie. 2015. "High-throughput phenotyping of cotton in multiple 29(6): 150-6.

Sharma, LK, H Bu, DW Franzen. 2016. "Use of corn height measured with an acoustic sensor improves yield estimation with ground based active optical sensors". *Computers and Electronics in Agriculture* 124(6): 254-62.

Shi, L, Shun Hu and YuanyuanZha. 2018."Estimation of sugarcane yield by assimilating UAV and ground measurements *via* ensemble Kalman filter. IEEE 8816-8819.

Shi, Y, JA Thomasson, SC Murray, NA Pugh, WL Rooney, S Shafian, N Rajan, G Rouze, CLS Morgan, HL Neely. 2016. "Unmanned Aerial Vehicles for High-Throughput Phenotyping and Agronomic Research". *PLoS ONE* 11: e0159781.

Shi, Y, M Chen, X Wang. 2018. "Numerical simulation and optimization of scattering performance of a conical centrifugal variable rate fertilizer spreader". *International Journal of Agricultural and Biological Engineering* 27(1): 34-43.

Shi, Y, M Chen, XC Wang. 2018. "Design and experiment of variable-rate fertilizer spreader with centrifugal distribution cover for rice paddy surface fertilization". *Transactions Chinese Society of Agricultural Machinery* 49(3): 86-93.

Shi, Y, Z Hu, X Wang. 2018. "Fertilization strategy and application model using a centrifugal variable-rate fertilizer spreader". *International Journal of Agricultural and Biological Engineering* 11(6): 41-48.

Shi,Y, Yang Zhu, Xiaochan Wang, Xin Sun, Yangfen Ding, Wexing Cao and Zhichao Hu. 2020. "Progress and development on biological information of crop phenotype research applied to realtime variablerate fertilization. *Plant Methods* 16:11.

Shibayama, M and TA Akiyama. 1986. "A spectroradiometer for field use: VI. Radiometric estimation for chlorophyll index of rice canopy". *Japanese Journal of Crop Science* 55(4): 433-438.

Smart, AS, AR Weeks, AR van Rooyen, A Moore, MA McCarthy and R Tingley. 2016. "Assessing the cost-efficiency of environmental DNA sampling". *Methods in Ecology and Evolution* 7: 1291-1298. https://doi. org/10.1111/2041-210x.12598

Smart, AS, R Tingley, AR Weeks, AR van Rooyen and MA McCarthy. 2015. "Environmental DNA sampling is more sensitive than a traditional survey technique for detecting an aquatic invader". *Ecological Applications* 25: 1944-1952. https://doi.org/10.1890/14-1751.1

Smith, K. 2015. "The use of drones in environmental management". World Environmental and Water Resources Congress., Proceedings ASCE, 1352-1361. doi: 10.1061/978078 4479162.133.

Sobayo, R, HH Wu, R Ray and L Qian. 2018. "Integration of Convolutional Neural Network and Thermal Images into Soil Moisture Estimation". In Proceedings of the 2018 1st International Conference on Data Intelligence and Security (ICDIS), South Padre Island, TX, USA, 8–10 April 2018. 207-210.

Somard, J, Mohammad Dalower Hossain, SarawutNinsawat andVorraveerukornVeerachitt. 2018. "Pre-harvest Sugarcane Yield Estimation Using UAV-Based RGB Images and Ground Observation". Sugar Research & Promotion.

Song Xiu-Peng,Yong -Jian Liang, Xiao Qiu Zhang, Zhen-Qiang Qin, Jin-Ju Wei, Yang-Rui Li and Jian-Ming Wu. 2020. "Intrusion of Fall Armyworm (*Spodoptera frugiperda*) in Sugarcane and Its Control by Drone in China". https://doi.org/10.1007/s12355-020-00799-x

Song, X, J Wang, W Huang, G Yan. 2009. "Monitoring spatial variance of winter wheat growth and grain quality under variable-rate fertilization conditions by remote sensing data". *Transactions Chinese Society of Agricultural Engineering* 25(9): 155-62.

Stagakis, S, V Gonzalez-Dugo, P Cid, M Guillen-Climent and P Zarco-Tejada. 2012. "Monitoring Water Stress and Fruit Quality In An Orange Orchard Under Regulated Deficit Irrigation Using Narrow-band Structural and Physiological Remote Sensing Indices". *ISPRS Journal of Photogrammetry and Remote Sensing* 71: 47-61.

Starks, PJ, D Zhao, WA Phillips and SW Coleman. 2006. "Development of canopy re- flectance algorithms for real-time prediction of bermudagrass pasture biomass and nutritive values". *Crop Science* 46: 927-934.

Starks, PJ, MA Brown, KE Turner and BC Venuto. 2016."Canopy visible and near-infrared reflectance data to estimate alfalfa nutritive attributes before harvest". *Crop Science* 56: 484-494. http://dx.doi.org/10.2135/cropsci2015.03.0162

Stucky, JM. 1984. "Comparison of two methods of identifying weed seedlings". *Weed Science* 32: 598-602.

Sullivan, DG, JP Fulton, JN Shaw and G Bland. 2007. "Evaluating The Sensitivity of An Unmanned Thermal Infrared Aerial System to Detect Water Stress In A Cotton Canopy". *Proc. ASABE* 50: 1955-1962.

Sullivan, DV,J Zheng, BMLobitz, JG Leung, BGallmeyer, M Aoyagi, RESlye andJ Brass.2004. "Demonstration of UAV-based imaging for agricultural surveillance and decision support". *Computers and Electronics in Agriculture* 44: 49-61.

Sun, L, F Gao, MC Anderson. 2017. "Daily mapping of 30 m LAI and NDVI for Grape yield prediction in California Vineyards". *Remote Sensing* 9:317. doi: 10.3390/rs9040317

Swain, KC, HPW Jayasuriya and VM Salokhe. 2007. "Suitability of Low altitude Remote Sensing Images for Estimating Nitrogen Treatment Variations in Rice Cropping for Precision Agriculture Adoption". *Journal of Applied Remote Sensing* 1: 1-11.

Tang, H, J Wang and C Xu. 2019. "Research progress analysis on key technology of chemical fertilizer reduction and efficiency increase". *Transactions of the Chinese Society for Agricultural Machinery* 50(4): 1-19.

Tang, H. 2018. "Progress and prospect of agricultural remote sensing research". *Journal of Agricultural* 8(1): 167-71.

Tang, L, YD Li, YP Zhang. 2011. "Simulation of canopy light distribution and application in rice". *Chinese Journal of Rice Science* 25(4): 427-34.

Tang, Q, L Chen, R Zhang, M Xu, G Xu and T Yi. 2016. "Airspeed and pressure affect spray droplet spectrum from an aerial nozzle for fixed-wing applications". Proceedings of the 13th International Conference on Precision Agriculture July 31 – August 3, 2016, St. Louis, Missouri, USA.

Tanriverdi, C. 2006. "A review of remote sensing and vegetation indices in precision farming". *J. Sci. Eng.,*9: 69-76.

Tarca, AL, VJ Carey, XW Chen, R Romero and S Dr¢aghici. 2007. "Machine learning and its applications to biology." *PLOS Computational Biology* 3, e116.

Tetracam, Inc. ADC Users Guide V2.3; Tetracam, Inc.: Chatsworth, CA, USA, 2011.

Thenkabail, PS, JG Lyon and A Huete. 2011. "Hyperspectral Remote Sensing ofVegetation". Boca Raton: CRC Press, 1943-1961. doi: 10.1201/b11222.

Tian, M, S Ban, Q Chang. 2016. "Use of hyperspectral images from UAV based imaging spectroradiometer to estimate cotton leaf area index". *Transactions Chinese Society of Agricultural Engineering* 32(21): 102-8.

Torrent, X, C Garcerá, E Moltó, P Chueca, R Abad, C Grafulla, 2017. "Comparison between standard and drift reducing nozzles for pesticide application in citrus: Part I. Effects on wind tunnel and field spray drift". *Crop Protection* 96: 130-143.

Torres-Sánchez, J, AI de Castro, JM Peña, FM Jiménez-Brenes, O Arquero, M Lovera and F López-Granados. 2018. "Mapping the 3D structure of almond trees using UAV acquired photogrammetric point clouds and object-based image analysis". *Biosystems Engineering* 176: 172-184.

Tran, CD and VI Grishko.2007. "Determination of water contents in leaves by a near-infrared multispectral imaging technique". *Microchemical Journal* 76: 91-94.

Tsouros, DC, PN Smyrlis, MG Tsipouras, DG Tsalikakis, N Giannakeas, AT Tzallas and P Manousou. 2017. "Automated collagen proportional area extraction in liver biopsy images using a novel classification via clustering algorithm". In Proceedings of the 2017 IEEE 30th International Symposium on Computer-Based Medical Systems (CBMS), Thessaloniki, Greece, 22–24 June 2017; pp. 30–34.

Tucker, CJ, BN Holben, JHJr Elgin and JE Mcmurtrey. 1980. "Relationship ofspectral data to grain yield variation". *Photogrammetric Engineering and Remote Sensing* 46: 657-666.

Varela, S, P Dhodda, W Hsu, P Prasad, Y Assefa, N Peralta, T Griffin, A Sharda, A Ferguson and I Ciampitti. 2018. "Early-season stand count determination in corn *via* integration of imagery from unmanned aerial systems (UAS) and supervised learning techniques". *Remote Sensing* 10: 343.

Vega, FA, FC Ramírez, MP Saiz and FO Rosúa. 2015. "Multi-temporal imaging using an unmanned aerial vehicle for monitoring a sunflower crop". *Biosystems Engineering* 132: 19-27. doi: 10.1016/j.biosystemseng.2015.01.008

Vermeulen, C, P Lejeune, J Lisein, P Sawadogo and P Bouch. 2013. "Unmanned aerial survey of elephants". *PloS ONE* 8: e54700.

Verrelst, J, G Camps-Valls, J Muñoz-Marí, J PabloRiveraa, F Veroustraeteb, JGPW Cleversc. 2015. "Optical remote sensing and the retrieval of terrestrial vegetation bio-geophysical properties - a review". *ISPRS Journal Photogrammetry and Remote Sensing*. 108: 273-290.

Villero, D, M Pla, D Camps, J Ruiz-Olmo and L Brotons. 2017."Integrating species distribution modelling into decision-making to inform conservation actions". *Biodiversity and Conservation*26: 251-271.

Wahab, I, Ola Hall and Magnus Jirström. 2018."Remote Sensing of Yields: Application of UAV Imagery-Derived NDVI for Estimating Maize Vigor and Yields in Complex Farming Systems in Sub-Saharan Africa".*Drones*2: 28. doi:10.3390/drones2030028

Wan, L, Y Li, H Cen, J Zhu, W Yin, W Wu, H Zhu, D Sun, W Zhou and Y He. 2018. "Combining UAV-Based Vegetation Indices and Image Classification to Estimate Flower Number in Oilseed Rape". *Remote Sensing* 10: 1484.

Wang SL, JL Song, XK He, L Song, XN Wang, CL Wang. 2017. "Performances evaluation of four typical unmanned aerial vehicles used for pesticide application in China". *International Journal of Agriculture and Biological Engineering* 10(4): 22-31.

Wang, A, W Zhang, W., & and X Wei. 2019. "A review on weed detection using ground-based machine vision and image processing techniques". *Computers and Electronics in Agriculture* 158: 226-240.

Wang, CL, XK He and XN Wang. 2016. "Testing method of spatial pesticide spraying deposition quality balance for unmanned aerial vehicle". *Transactions of the Chinese Society for Agricultural Engineering* 32: 54-61.

Wang, J, YB Lan, HH Zhang, YL Zhang, S Wen, WX Yao. 2018. "Drift and deposition of pesticide applied by UAV on pineapple plants under different meteorological conditions". *International Journal of Agriculture and Biological Engineering* 11(6): 5-12.

Wang, J, YB Lan, HH Zhang, YL Zhang, S Wen, WX Yao. 2018. "Drift and deposition of pesticide applied by UAV on pineapple plants under different meteorological conditions". *International Journal of Agriculture and Biological Engineering* 11(6): 5-12. DOI: 10.25165/j.ijabe.20181106.4038

Wang, J, ZJ Wang, WJ Huang. 2004. "The vertical distribution characteristic and spectral response of canopy nitrogen in different layer of winter wheat". *International journal of remote sensing* 8(4): 309-16.

Wang, JH, WJ Huang, CL Lao. 2007. "Inversion of winter wheat foliage vertical distribution based on canopy reflected spectrum by partial least squares regression method". *Spectroscopy and Spectral Analysis* 27(7): 1319-22.

Wang, JJ, H Ge, Q Dai, I Ahmad, Q Dai, G Zhou, M Qin and C Gu. 2018. "Unsupervised discrimination between lodged and non-lodged winter wheat: A case study using a low-cost unmanned aerial vehicle". *International journal of remote sensing* 39: 2079-2088.

Wang, SL, JL Song and XK He. 2017. "Performances evaluation of four typical unmanned aerial vehicles used for pesticide application in China". *International Journal of Agriculture and Biological Engineering* 10: 22-31.

Wang, X, W Yang, A Wheaton, N Cooley and B Moran. 2010. "Automated canopy temperature estimation via infrared thermography: A first step towards automated plant water stress monitoring". *Computers and Electronics in Agriculture* 73(1): 74-83.

Wang, XC, M Chen, GX Sun. 2015. "Design and test of control system on variable fertilizer applicator for winter wheat". *Transactions Chinese Society of Agricultural Engineering* 31(Supp. 2): 88-92.

Weissensteiner, MH, JW Poelstra and JBW Wolf. 2015."Low-budget ready-to-fly unmanned aerial vehicles: An effective tool for evaluating the nesting status of canopy-breeding bird species". *Journal of Avian Biology*46: 425-430.

Weixiang Yao, Yubin Lan, W Clint Ho_mann, Jiyu Li, Shuang Guo, Huihui Zhang and Juan Wang. 2020. "Droplet size distribution characteristics of aerial nozzles by bell206l4 helicopter under medium and low airflow velocity wind tunnel conditions and field verification test". *Applied Sci*ence 10: 2179. doi:10.3390/app10062179

Westoby, MJ, J Brasington, NF Glasser, MJ Hambrey and J Reynolds. 2012. "Structure-from-Motionphotogrammetry: A low-cost, effective tool for geoscience applications". *Geomorphology* 179: 300-314.

Wiegand, C, A Richardson, D Escobar and A Gerbermann. 1991. "Vegetation indices in crop assessments". *Remote Sensing of Environment* 35: 105-119.

Wilke, N, Bastian Siegmann, Lasse Klingbeil, Andreas Burkart, Thorsten Kraska, Onno Muller, Anna van Doorn, Sascha Heinemann and Uwe Rascher. 2019."Quantifying Lodging Percentage and Lodging Severity Using a UAV-Based Canopy Height Model Combined with an Objective Threshold Approach". Remote Sensing 11: 515. doi:10.3390/rs11050515

Wilson, AM,J Barr and M Zagorski. 2017. "The feasibility of counting songbirds using unmanned aerial vehicles". *Auk*134: 350-362.

Woolley, JT. 1971. "Reflectance and transmittance of light by leaves". Plant Physiology 47: 656-662. doi: 10.1104/pp.47.5.656

Wu, C, W Li, Z Niu. 2010. "Nondestructive estimation of canopy chlorophyll content using Hyperion and Landsat/TM images". *International journal of remote sensing* 31(8): 2159-67.

Wu, C, X Han, Z Niu. 2010. "An evaluation of EO-1 hyperspectral Hyperion data for chlorophyll content and leaf area index estimation". *International journal of remote sensing* 31(4): 1079-86.

Wu, Q, B Qi, TJ Zhao. 2013. "A tentative study on utilization of canopy hyperspectral reflectance to estimate canopy growth and seed yield in soybean". *Acta Agronomica Sinica* 39(2): 309-18.

Xiao, Q, Rui Du, Lin Yang, Xiaoqiang Han, Sifeng Zhao, Guoqiang Zhang, Wei Fu, Guobin Wang and Yubin Lan. 2020. "Comparison of Droplet Deposition Control Efficacy on Phytophthora capsica and Aphids in the Processing Pepper Field of the Unmanned Aerial Vehicle and Knapsack Sprayer". *Agronomy* 10: 215. doi:10.3390/agronomy10020215

Xiao-Qiu Zhang, Xiu-Peng Song, Yong-Jian Liang, Zhen-Qiang Qin, Bao-Qing Zhang, Jin-Ju Wei, Yang-Rui Li and Jian-Ming Wu. 2020. "Effects of Spray Parameters of Drone on the Droplet Deposition in Sugarcane Canopy". https://doi.org/10.1007/s12355-019-00792-z

Xiao-Qiu Zhang, Yong-Jian Liang, Zhen-Qiang Qin, De-Wei Li, Chun-Yan Wei, Jin-Ju Wei, Yang-Rui Li and Xiu-Peng Song. 2019. "Application of Multi-rotor Unmanned Aerial Vehicle Application in Management of Stem Borer (Lepidoptera) in Sugarcane". https://doi.org/10.1007/s12355-018-0695-y

Xinyu Xue, Yubin Lan, Zhu Sun, Chun Chang and W Clint Hoffmann. 2016. "Develop an unmanned aerial vehicle based automatic aerial spraying system". *Computers and Electronics in Agriculture* 128: 58-66. http://dx.doi.org/10.1016/j.compag.2016.07.022

Xuan et al., 2019. DOI: 10.13031/aim.201901493

Xue, J and B Su. 2017. "Significant remote sensing vegetation indices: A review of developments and applications". *Journal of sensors* 1353691.

Xue, L and L Yang. 2008. "Recommendations for nitrogen fertiliser topdressing rates in rice using canopy reflectance spectra". *Biosystems Engineering* 100(4): 524-34.

Xue, X and Y Lan. 2013. "Agricultural aviation applications in USA". *Transactions of the Chinese Society for Agricultural Machinery* 44 (5): 194-201.

Xue, XY, K Tu, WC Qin, YB Lan and HH Zhang. 2014. "Drift and deposition of ultra-low altitude and low volume application in paddy field". *International Journal of Agriculture and Biological Engineering* 7(4): 23-28. DOI. 10.3965/j.ijabe.20140704.003

Xue, Z, H Gao and J Liu. 2011. "Monitoring growth and grain yield of wheat in fields with different soil fertility levels and different fertilizer application using spectral reflectance technique". *Journal of Triticeae Crops* 31(2): 324-330.

Yallappa, D, M Veerangouda, Devanand Maski, Vijayakumar Palled and M Bheemanna. 2018. "Development and evaluation of drone mounted sprayer for pesticide applications to crops". DOI: 10.1109/GHTC.2017.8239330

Yanbo Huang, W Clint Hoffman, Yubin Lan, K. Fritz Bradley and J. Thomson Steven. 2014. "Development of a Low-Volume Sprayer for an Unmanned Helicopter". *Journal of Agricultural Science* 7(1): 1916-9752. doi:10.5539/jas.v7n1p148

Yang, FB, XY Xue, L Zhang and Z Sun. 2017. "Numerical simulation and experimental verification on downwash air flow of six-rotor agricultural unmanned aerial vehicle in hover". *International Journal of Agriculture and Biological Engineering* 10(4): 41-53. DOI. 10.25165/j.ijabe.20171004.3077

Yang, G, C Zhao, Q Liu. 2011. "Inversion of a radiative transfer model for estimating forest LAI from multisource and multi-angular optical remote sensing data". *IEEE Transaction on Geoscience and Remote Sensing* 49(3): 988-1000.

Yang, JN, JC Zhang, X Yao. 2013. "Experiments on performance of portable plant growth monitoring diagnostic instrument". *Transactions Chinese Society of Agricultural Machinery* 44(4): 208-212, 201.

Yanhua Meng, Jianli Song, Yubin Lan, Guiying Mei, Zijing Liang and Yuxing Han. 2019. "Harvest aids efficacy applied by unmanned aerial vehicles on cotton crop. Industrial Crops & Products" 140: 111645. https://doi.org/10.1016/j.indcrop.2019.111645

Yano, IH, JR Alves, WE Santiago, and BJ Mederos. 2016. "Identification of Weeds In Sugarcane Fields Through Images Taken By UAV and Random forest Classifier". *IFAC-Paper son Line* 49: 415-420.

Yebra, M and E Chuvieco. 2009. "Linking ecological information and radiative transfer models to estimate fuel moisture content in the Mediterranean region of Spain: solving the ill-posed inverse problem". *Remote Sensing of Environment* 113(11): 2403-11.

Yinyan, S, C Man, W Xiaochan. 2018. "Numerical simulation of spreading performance and distribution pattern of centrifugal variable-rate fertilizer applicator based on DEM software". *Computers and Electronics in Agriculture* 144(1): 249-59.

Yinyan, S, H Zhichao, W Xiaochan. 2018. "Motion analysis and system response of fertilizer feed apparatus for paddy variable-Rate fertilizer spreader". *Computers and Electronics in Agriculture* 153(10): 239-47.

Yonah, IB, SK Mourice, SD Tumbo, BP Mbilinyi and J Dempewolf. 2018 "Unmanned aerial vehicle-based remote sensing in monitoring smallholder, heterogeneous crop fields in Tanzania". *International journal of remote sensing* 39: 5453-5471.

Yu, H, D Liu, G Chen. 2010. "A neural network ensemble method for precision fertilization modeling". *Mathematical and Computer Modelling* 51(11–12): 1375-1382.

Yuan, J, CL Liu and YM Li. 2010. "Gaussian processes based bivariate control parameters optimization of variable-rate granular fertilizer applicator". *Computers and Electronics Agriculture* 70(1): 33-41.

Yuan, J, QH Liu and XM Liu. 2014. "Granular multi-flows fertilization process simulation and tube structure optimization in nutrient proportion of variable rate fertilization". *Transactions of the Chinese Society for Agricultural Machinery* 45(11): 81-87.

Yuan, Y, S Li, X Fang. 2013. "Decision support system of N, P and K ratio Fertilization". *Transactions of the Chinese Society of Agricultural Machinery* 44(8): 240-244.

Yue, J, T Lei, C Li and J Zhu. 2012. "The application of unmanned aerial vehicle remote sensing in quickly monitoring crop pests". *Intelligent Automation and Soft Computing* 18: 1043-1052.

Zaman, QU, TJ Esau and AW Schumann. 2011. "Development of prototype automated variable rate sprayer for real-time spot-application of agrochemicals in wild blueberry fields". *Computers and Electronics in Agriculture* 76(2): 175-182.

Zang, Y, Y Zang, ZY Zhou, XY Gu, R Jiang, LX Kong. 2019. "Design and anti-sway performance testing of pesticide tanks in spraying UAVs". *International Journal of Agriculture and Biological Engineering* 12(1): 10-16. DOI: 10.25165/j.ijabe.20191201.4338

Zarco-Tejada, PJ, V Gonzalez-Dugo and JAJ Berni. 2012. "Fluorescence, temperature and narrowband indices acquired from a UAV platform for water stress detection using a micro-hyperspectral imager and a thermal camera". *Remote Sensing of Environment* 117: 322-337

Zermas, D, D Teng, P Stanitsas, M Bazakos, D Kaiser, V Morellas, D Mulla, N Papanikolopoulos. 2015. "Automation Solutions for the Evaluation of Plant Health in Corn Fields". In Proceedings of the 2015 IEEE/RSJ International

Conference on Intelligent Robots and Systems (IROS), Hamburg, Germany, 28 September- 2October 2015; pp. 6521-6527.

Zhang S C, Xue X Y, Sun Z, Zhou L X, Jin Y K. 2017. "Downwash distribution of single-rotor unmanned agricultural helicopter on hovering state". *International Journal of Agriculture and Biological Engineering* 10(5): 14-24. DOI: 10.25165/j. ijabe.20171005.3079

Zhang, C and JM Kovacs. 2012. "The application of small unmanned aerial vehicle systems for precision agriculture: a review". *Precision Agriculture* 13: 693-712.

Zhang, DY, X Wang, C Coburn. 2013. "Design and experiment of ground based agriculture-oriented multi-angle observation device". *Transactions Chinese Society of Agricultural Machinery* 44(1): 174-178.

Zhang, DY, YB Lan, LP Chen, X Wang and D Liang.2014. "Current status and future trends of agricultural aerial spraying technology in China". *Transactions of the Chinese Society for Agricultural Machinery* 45: 53-59.

Zhang, J, Chufeng Wang, Chenghai Yang, Tianjin Xie, Zhao Jiang, Tao Hu, Zhibang Luo, Guangsheng Zhou and Jing Xie. 2020. "Assessing the Effect of Real Spatial Resolution of InSitu UAV Multispectral Images on Seedling Rapeseed Growth Monitoring". *Remote Sensing* 12: 1207. doi:10.3390/rs12071207.

Zhang, J, XK He, LJ Song, AJ Zeng, YJ Liu and XF Li. 2012. "Influence of spraying parameters of unmanned aircraft on droplets deposition". *Transactions of the Chinese Society of Agricultural Machinery* 43(12): 94-96.

Zhang, R. 2012. "The research on key technology of intelligence variable rate Fertilization". Bejing: Chinese Academy of Agricultural Sciences.

Zhang, SC, XY Xue, WC Qin, Z Sun, SM Ding S M and LX Zhou. 2015. "Simulation and experimental verification of aerial spraying drift on N-3 unmanned spraying helicopter". *Transactions of the Chinese Society of Agricultural Engineering* 31(3): 87-93.

Zhang, X, Y Jinming Zhao, Guijun Yang, Jiangang Liu, Jiqiu Cao, Chunyan Li, Xiaoqing Zhao and JunyiGai. 2019. "Establishment of Plot-Yield Prediction Models in Soybean Breeding Programs Using UAV-Based Hyperspectral Remote Sensing". *Remote Sensing* 11: 2752. doi:10.3390/rs11232752

Zhang, Y, X Yao, YC Tian. 2010. "Estimating leaf nitrogen content with near infrared reflectance spectroscopy in rice". *Chinese Journal of the Plant Ecology* 34(6): 704-12.

Zhao, C. 2020. "Research and practice of precision agriculture". Beijing: Science Press; 2009. Shi. *Plant Methods* 16(11): 13-15

Zhao, T, B Stark, Y Chen, AL Ray and D Doll. 2017. "Challenges In Water Stress Quantification Using Small Unmanned Aerial System (sUAS): Lessons From A Growing Season of Almond". *Journal of Intelligent and Robotic Systems* 88: 721-735.

Zhao, T, D Doll, D Wang and Y Chen. 2017. "A New Framework for UAV-based Remote Sensing Data Processing and Its Application in Almond Water Stress Quantification". In Proceedings of the 2017 International Conference on Unmanned Aircraft Systems (ICUAS), Miami, FL, USA, 13-16 June 2017; pp. 1794-1799.

Zhaoxia Lou, Fang Xin, Xiaoqiang Han, Yubin Lan, Tianzhu Duan and Wei Fu. 2018. "Effect of Unmanned Aerial Vehicle Flight Height on Droplet Distribution, Drift and Control of Cotton Aphids and Spider Mites". *Agronomy* 8: 187 doi:10.3390/agronomy8090187

Zheng, H, T Cheng, D Li. 2018. "Combining unmanned aerial vehicle (UAV)-based multispectral imagery and ground-based hyperspectral data for plant nitrogen concentration estimation in rice. *Frontiers in Plant Science* 9(7): 936-49.

Zheng, H, W Li, J Jiang, Y Liu, T Cheng, Y Tian, Y Zhu, W Cao, Y Zhang and X Yao. 2018. "A Comparative Assessment of Different Modeling Algorithms for Estimating Leaf Nitrogen Content in Winter Wheat Using Multispectral Images from an Unmanned Aerial Vehicle". *Remote Sensing* 10: 2026.

Zheng, YJ, SH Yang, and CJ Zhao. 2017. "Modelling operation parameters of UAV on spray effects at different growth stages of corns". *International Journal of Agriculture and Biological Engineering* 10: 57-66.

Zhenhua, X, G Caijuan and M Wenqi. 2011. "Study on the relationship among yield, nutrient efficiency and benefits of grain crop in typical regional areas of China". *Chinese Agricultural Science Bulletin* 27(11): 116-122.

Zhu, H, M Salyani and RD Fox. 2011. "A portable scanning system for evaluation of spray deposit distribution". *Computers & Electronics in Agriculture* 76(1): 38-43.

Zhu, Y, Y Zhu, Y Huang. 2010. "Assimilation technique of remote sensing information and rice growth model based on particle swarm optimization". *Journal of remote sensing* 14(6): 1233-1249.

Zhu,J, K Wang, J Deng and T Harmon. 2009. "Quantifying Nitrogen Status of Rice Using Low Altitude UAV-Mounted System and Object-Oriented Segmentation Methodology". In Proceedings of the International Design Engineering Technical Conferences and Computers and Information in Engineering Conference, San Diego, CA, USA, 30 August-2 September 2009; pp.1- 7.

Ziliani, M, S Parkes, I Hoteit and M McCabe. 2018. "Intra-Season Crop Height Variability at Commercial Farm Scales Using a Fixed-Wing UAV". *Remote Sensing 10: 2007.*

Index